THE
BITTER TRUTH
ABOUT
ARTIFICIAL
SWEETENERS

THE *BITTER TRUTH* ABOUT ARTIFICIAL SWEETENERS

Dennis W. Remington, M.D.
Barbara W. Higa, R.D.

VITALITY
HOUSE
INTERNATIONAL
INC.

Copyright ©1987 by
Vitality House International, Inc.
3707 North Canyon Road #8-C
Provo, Utah 84604

Telephone: 801-224-9214

First Printing, September, 1987

This collection of recipes has been gathered and adapted from a variety
of sources.

Library of Congress Catalog Card Number 87-50668

ISBN 0-912547-04-9

Printed in the United States of America

Table of Contents

List of Diagrams

List of Tables

About the Authors

Dennis W. Remington, M.D., is a family physician in private practice with an interest in nutrition, exercise, and food and chemical allergies. He has developed several innovative concepts in weight management. He is director of the Eating Disorder Clinic at Brigham Young University's Student Health Center in Provo, Utah, a member of the American Society of Bariatric Physicians, and the American Academy of Environmental Medicine. He was the founding president of the Society for the Study of Biochemical Intolerance. He is a coauthor of *How to Lower Your Fat Thermostat, Back to Health - A Comprehensive Medical and Nutritional Yeast Control Program*, and the audiotape program, *The Neuropsychology of Weight Control*.

Barbara W. Higa, Registered Dietitian, works in a medical practice with patients suffering from food and chemical allergies, obesity, and yeast problems. A graduate of Brigham Young University, she completed her dietetic internship at the Veterans Administration Hospital in Los Angeles, and taught quantity food production at Brigham Young University. She has instructed physicians from the United States, Canada, and Australia on new allergy treatment and yeast treatment techniques. She has taught aerobic exercise, and specializes in developing delicious recipes for healthy eating. She is a coauthor of *Back to Health - A Comprehensive Medical and Nutritional Yeast Control Program*.

An Importance Notice

This book was written as a source of information for professionals and for lay people. Although the dietary advice given is similar to that recommended by most knowledgeable health professionals, anyone with health problems should seek the advise of their own personal physician before making any significant dietary changes or before starting an exercise program.

Acknowledgements

We would like to acknowledge the scientific work of Dr. Richard Wurtman in the area of artificial sweeteners and his encouragement in writing this book. We would also like to thank Dr. Woodrow Monte for his work and encouragement.

Many thanks to Kathy Frandsen for her editing skills, Rick Thayne for his talented work with the cover design and graphics, Mitch Stowell for his computer expertise, and Karen Murdock for her encouragement in creating the recipes.

We are grateful for the many patients who have shared their experiences with us regarding artificial sweetener usage.

We would also like to acknowledge all of the many scientists whose work has helped to make this book possible.

THE BITTER TRUTH - FOREWORD

We are all bombarded by advertisements claiming benefits and pleasures from using artificially sweetened products. The claims implied by these advertisements are often exaggerated, not supportable by scientific research, and sometimes blatantly dishonest. In this book, we present another side of artificial sweeteners and dieting. We question the basic, fundamental concept of simple energy balance in relation to weight gain that has been taught as indisputable truth for many years. We also question the entire concept of reducing caloric intake for weight control.

The information we present has been obtained from clinical observations, patient histories, and from hundreds of scientific research articles. In some cases, there are conflicting findings in the scientific literature. We have made no attempt to present all of the conflicting data. Some of the theories we present have inadequate scientific evidence to support them at the present time, but are based on consistent clinical observations and indirect evidence. We believe that these theories will eventually be substantiated through future research and as more clinical experience is gained with the new artificial sweeteners.

We have no economic interest or ties with any food or drink manufacturers, nor do we obtain any research funds from any such company. We work in a fee-for-service medical practice. Our motivation for writing this book is based on an urgent need for people to become informed about the effects of artificial sweeteners and restrictive dieting.

We are appalled at the tremendous amount of outdated, incorrect information that continues to be presented by scientific "authorities" from respected universities and research institutes. This misinformation places a tremendous burden of guilt on overweight people—guilt which is entirely unjustified. It also causes them to be treated unfairly, and with undeserved contempt. Even worse, believing these incorrect concepts, overweight people and those afraid of becoming overweight restrict food intake, and produce suffering and grief for themselves. They not only become fatter, but create various physical and emotional health problems as well.

It is our hope that people everywhere will eventually understand correct scientific principles related to weight control, and start using sensible methods that will improve, instead of damage their health and well-being.

INTRODUCTION

THE BITTER TRUTH ABOUT ARTIFICIAL SWEETENERS

What do you know about artificial sweeteners?

Think about the messages you're bombarded with daily.

Every night on television you can see scantily clad young men and women frolicking in endless summer sunshine, a diet drink in one hand and an attractive companion in the other.

The message?

Diet drinks help you to have fun and be thin!

If you've fallen for it, you're not alone. In response to this advertising blitz, people are using artificially sweetened products in huge quantities—for snacks, with meals, and even as a substitute for breakfast. Most people use them in an attempt to control weight, and many of the commercial weight-loss organizations promote them for the same purpose.

A recent survey by the American Cancer Society on 78,000 Americans revealed that those regularly using artificial sweeteners had a higher incidence of weight gain than those who didn't.

What is the truth? Who can you believe? Will artificial sweeteners make your body irresistibly thin, or will they transform you into a mound of fat?

After becoming the most intensively researched food additive in history, the artificial sweetener **aspartame** (NutraSweet® and Equal®) was approved in 1981 by the FDA. In spite of reassurances from them that it is completely safe, concerned scientists and consumer groups have collected more than ten thousand formal complaints, and claim that there are tens of thousands of other people damaged by its use. A support group named **Aspartame Victims and Their Friends** has even been organized. The "friends" hope to help those "victims" who have been blinded or otherwise incapacitated after using aspartame.

What is the truth? Are artificial sweeteners completely safe for everyone in unlimited quantities? Are there dangers and side effects you should know about? What about the cancer issue? Can artificial sweeteners help you eat a healthier diet, or will they only make you unhealthy?

You'll find the answers to these questions in the pages of this book—in addition to important information about sugar and caffeine. If you are using artificially sweetened products, or thinking about using them, you should know the issues.

Ready?

Let's start with the weight issue. . . .

SECTION I

CHAPTER ONE

DO ARTIFICIAL SWEETENERS HELP REDUCE CALORIES AND CONTROL WEIGHT?

Have you been using artificial sweeteners to save calories—and to help you lose or control your weight?

If so, you're in for a big surprise.

Why?

Because scientific evidence now strongly suggests that just the opposite might happen!

Under some circumstances, experimental animals given artificial sweetener in their drinking water ate more and became fatter.[1,2,3,4] Only when the artificial sweetener was given in massive enough doses to make the food taste bad did the animals eat less over the long run.[5,6]

From the thousands of patients we have treated, it seems that some people can effectively use artificial sweeteners to help reduce calories as long as they are on a calorie-restricted diet. A diet drink with few or no calories may help temporarily relieve hunger by partly filling the stomach, and it might give some partial satisfaction because of the sweetness. Artificial sweeteners might help to cut calories, then, if you continue to make a conscious effort to ignore your hunger drive and eat less.

But how long can you keep that up?

Is it helpful in the long term to reduce calories?

If you've ever tried to stick to a calorie-restricted diet for a prolonged period of time and found it was too tough to do, there are good reasons why.

Compelling evidence shows that the longer people restrict food intake and attempt to diet, the hungrier they become and the more difficult it is to stay with the diet. Eventually, almost everyone attempting a calorie-restricted diet will give in to increasing hunger and will go off the diet.

Rather than going on a formal calorie-restricted diet, many people who want to lose weight simply try little tricks to reduce their day-to-day caloric intake, but don't limit themselves to a certain number of calories. Some do this by drinking artificially sweetened drinks in place of a meal or a snack, or by using other reduced-calorie products in place of real food. What happens? They often eat just as much food over the long run.

But what if you are successful in making a conscious effort to restrict yourself to a certain number of calories? Can you lose weight permanently simply by reducing calories? We used to think so, but now we know that the process isn't that simple.

All of us have seen short-term evidence that if you eat fewer calories, you can lose weight. Maybe you've really binged on rich foods during a Christmas holiday season, and have seen the scales creep up as a result. Then, when you've pared your calories to a minimum for a few weeks afterward, you have probably been able to lose the weight you gained. But remember— we're talking a few weeks! That's the problem: most studies on weight management have concentrated only on short-term observations. Many studies were done by university students as part of research projects—and as a result, they lasted only a few weeks, or, at most, several months. The few studies that have concentrated on a longer-term observation are loaded with surprises that change the way we must think about calories and weight loss.

In the 1975 HANES study (a survey done to evaluate nutritional status), the caloric intake and body weight of thousands of Americans were assessed.[7] People were classified as being underweight, normal weight, or overweight. The overweight were grouped according to how overweight they were. There was a very clear relationship between body weight and calories—but it's not the relationship you'd expect! What did researchers find? Here's one of the surprises: those who were thin ate considerably more than those who were fat! Another surprise? The fatter the people were the less food they ate! (See Fig. 1.)

The heaviest group included massively obese people. This group ate more than some of the other overweight groups in order to sustain their large muscle mass (carrying that much weight around is like doing weight lifting all the time), but they still ate *less* than the people who were thin! Even with huge muscles and increased demand for energy, they ate considerably less per pound of body weight than did thin people.

Many similar studies have shown that overweight people eat less than—or at least no differently from—normal weight people.[8,9,10,11,12]

Some interesting studies have been done with overweight and normal weight children in various controlled circumstances. In a school situation, obese girls ate about 1965 calories per day, while their lean classmates ate 2706 calories.[13] During a controlled camp study, the thin boys ate an average of 4628, while the obese group ate 3430 calories.[14] In these situations, the obese children were noted to be less active than the normal weight children—but the difference in activity didn't even begin to account for the tremendous difference in caloric intake. With so many studies challenging our old ideas about calories and weight problems, it's time to take a fresh look at some old, well-accepted traditional concepts.

Analysis of Simple Energy Balance Theory Of Obesity

The traditional view says that your body is like a passive tank of fat—and that the amount of fat it contains equals the calories you eat minus the calories you burn up. (See Fig. 2.) According to this theory, if you take in more calories than you can burn up, you get fat. Also according to this theory, if you simply take in fewer calories or burn up more of them, you'll automatically get thin.

This whole concept seems so obvious that for literally hundreds of years no one challenged it. No one made an attempt to evaluate it. But if weight were really controlled by such a simple

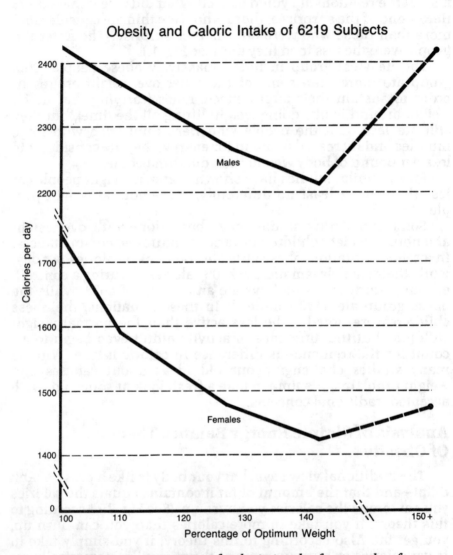

Figure 1. Comparison of caloric intake for people of various body weight. Note that in both men and women, the thinnest people ate the most, and up to 149 percent above ideal weight, the more overweight people were, the less they ate. Drawn from data presented in Hanes 1 study.[7]

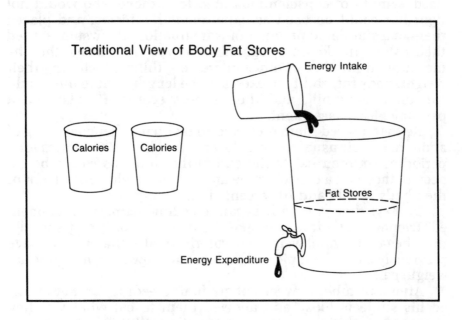

Figure 2. Body portrayed as passive tank for the storage of fat. The amount of fat stored was thought to be simply the amount of energy taken in by eating, minus the energy expended through exercise and the metabolic needs of the body.

energy balance concept, it should be very easy to evaluate that relationship.

If the simple energy balance theory were correct, it should be easy for a group of experimental subjects to simply reduce the number of calories they take in and, as a result, to lose weight gradually and steadily. It should then be very simple to maintain that weight level merely by eating slightly fewer calories than they did before they started—since their new streamlined bodies would require fewer calories than before.

But it simply doesn't work that way.

Perhaps the best controlled human study ever reported concerning this was performed by Dr. Keys during World War II.[15] He used a group of conscientious objectors, those who would not fight but could be used as experimental subjects, and placed them in a simulated prisoner-of-war situation. He first measured their caloric intake on a regular eat-as-you-want basis; then he measured a number of other interesting things, including their weight, body fat, ability to exercise, the length of time they could run on a treadmill, and their fitness scores. He also did a psychological assessment.

He then placed them on a calorie-restricted diet consisting of about half their usual intake. They ate roughly 1700 calories for a period of six months. At the end of that time, as would be expected, they lost a considerable amount of weight—25 percent of their body weight and 70 percent of their body fat.

Dr. Keys then gradually began refeeding them. What happened? Even when their food intake was increased only very slightly, they began to gain weight. A caloric intake that would have previously caused a rapid weight loss was now causing a gradual weight *gain*.

After a number of weeks of gradual refeeding, Dr. Keys turned his subjects loose and allowed them to eat whatever they chose. Then things really got interesting. Within a few weeks they had returned to their original weight. But that's not all! By the thirty-third week after refeeding began, they were heavier than they were at the beginning of the study, and carried considerably more body fat! (See Fig. 3.) Even though they ate a few more calories during refeeding than they had eaten before beginning the test, the difference was not nearly enough to explain the rapid weight gain.

Let's look at the other end of the spectrum now. If the simple energy balance theory of weight control is correct, then it should be possible to give people extra food and watch them gain weight steadily and predictably. But it doesn't happen. A number of studies have been attempted in which young university students were given access to plenty of high-calorie foods and were encouraged to eat as much as they could. In some of these studies, some subjects ate *thousands* of extra calories per day, with no significant weight gain at all![16] In other situations, those who were force-fed up to *8,000 calories* a day did gain some weight, but quickly returned to their normal weight as soon as they resumed normal eating.[17] In order to maintain the higher weight,

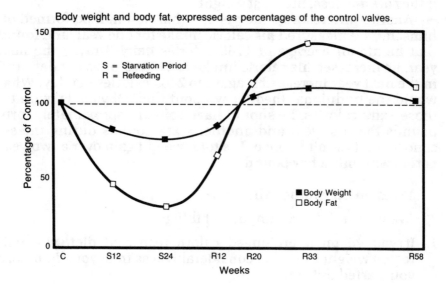

Body weight and body fat, expressed as percentages of the control valves.

Figure 3. Keys starvation study. After 24 weeks of reduced food intake (S for starvation period), note that body weight had fallen by about 25 percent and that fat stores had fallen by about 70 percent. During refeeding period, weight was regained rather slowly at first since food intake was increased gradually. Note rapid increase in body fat with free eating after 12 weeks (R12). After 33 weeks of refeeding (R33), the body fat was about 40 percent greater than at the beginning of the study, while body weight was only 10 percent higher. Note that after more than one year of refeeding, body fat was still about 10 percent higher than at the beginning, even though body weight was only slightly higher.[15]

some subjects had to eat 2,400 calories a day more than usual[18]; if they ate any less, they lost weight.

Another interesting study is one that Neuman performed on himself.[19] He measured his caloric intake for one year, and found that he ate an average of 1766 calories daily. During the next year, he increased his caloric intake to 2199 calories per day, and in the next year increased it again to 2403 calories per day. What would the traditional theory say about this? Obviously, with all those extra calories, he should have gained approximately forty pounds the first year and another sixty pounds during the second year. It didn't happen. His total weight gain over a two-year period was only a few pounds.

What do the studies show us?

We've learned three important things:

1. If you go on a prolonged calorie-restricted diet, you will regain weight by eating considerably less than you did before you started the diet.

2. If you force-feed and gain weight, you will lose it again even though you are eating considerably more calories than you previously ate to maintain your stable weight.

3. A scientist maintained his weight within a few pounds even though he ate considerably more food.

These observations suggest that there really isn't a good relationship between the number of calories you eat and your long-term weight. We want to make it very clear that we are not trying to deny any basic laws of physics and thermodynamics. We are not even saying that calories do not have some influence on body weight. All we are saying is that there are many factors (in addition to the number of calories you eat and the number you burn up) that control body weight and fat storage.

Let's look at some more studies that begin to give some answers. All of the following studies show that you can increase body fat without increasing the number of calories eaten. In every study, a control group given exactly the same number of calories remained thin, while the studied subjects got fat.

Study Situations

1.Eating one daily meal

One group of experimental rats was given all the food they desired, readily available at all times throughout the day. The amount of food they ate was measured, and exactly the same amount of food was given to a second group. Rats in the second group, however, were required to eat all of their food during a one-hour time period, and then went twenty-three hours with no food. The rats that ate only one meal a day became much fatter.[20,21,22]

2. Insulin injections

Two groups of rats were fed exactly the same number of calories and exercised exactly the same way. The difference? One group was given injections of insulin. What happened? The rats that were given insulin became fatter. (When given insulin injections, rats or any other mammal normally increase their food intake and gain weight very rapidly.)[23,24] Even when the experiment is designed to limit the food intake to the same amount eaten by normal rats, there is excessive fat gain, although it is slower and not as extensive as when the rats are given free access to food.

3. Inactivity

If two groups of animals have different activity levels, the inactive group becomes fatter than the active group. Then that proves the energy balance theory, right? Wrong. When the active rats are given more food to compensate for the energy they are burning up, they are still more lean than the less active rats. There is something about exercise which causes animals to maintain less body fat than would be expected through looking only at the expenditure of energy.[25,26,27,28]

4. Refined sugar

Experimental animals given access to refined sugar will eat more food, and will become very fat.[29,30,31,32] Even when two groups of animals eat exactly the same number of calories, the group using refined sugar will become fatter than the control group.[33]

5. Dietary fat

Two groups of animals that are fed exactly the same number of calories should theoretically weigh the same. But if one group gets foods high in fat and the other gets foods low in fat, the group getting the high-fat food becomes considerably fatter—even though the *same* number of calories are ingested by both groups![34,35,36]

6. Genetic obesity

Some strains of experimental animals have an interesting genetic pattern. The Zucker rat has an inherited pattern that causes 50 percent of the rats in a litter to inherit a trait which causes them to become fat, while the other 50 percent will be normal weight. The rats with the fat trait generally will eat considerably more food and gain much more weight. But the key isn't necessarily too much food: when they are given only the same amount of food that their lean litter mates eat, they will still become significantly fatter than the thin group, even though they don't gain as much as they would if given free access to food.[37]

7. Brain damage

Brain damage can be inflicted on rats by making surgical incisions in the hypothalamus area of the brain. These animals tend to eat more food, and become fatter. Even when their food intake is restricted to the same level as the control rats, the increased weight still occurs.[38,39,40]

By injecting materials that are toxic to certain areas of the brain, obesity can be produced without any increase in food intake.[41,42] In one study, brain-damaged rats actually decreased their intake of highly palatable food, and increased their body fat by 36 percent.[43]

8. Morphine injections

In another study, rats who were given morphine ate voraciously, but experienced a decrease in body weight.[44]

These studies—as well as a great number of similar ones—should prove that there isn't a clear-cut relationship between how much you eat and how much you weigh. A recent research article discussed 12 types of experimental obesity, and pointed out that in 11 of the 12, obesity could be produced without any extra eating.[45] Weight gain can occur with no increase in food intake, and weight loss can occur with no decrease in food intake. Weight gain can even occur in spite of eating less food, and weight loss can occur in spite of eating more food.

Do Artificial Sweeteners Help Control Weight?

So where do artificial sweeteners fit in?

During the years we have worked with thousands of overweight people, we've noticed that those who use artificial sweeteners seem to have a more difficult time losing weight than those who do not. In searching the literature to find out why, a number of articles have been found suggesting that artificial sweeteners are associated with weight gain, both in experimental animals and in people.

One of the authors of this book (DWR), was recently invited to participate in a panel discussion on the benefits of aspartame, sponsored by a group of diet product manufacturers. Two prominent research scientists spoke in favor of aspartame, both of whom had been paid by a manufacturer to do experimental work with aspartame. Dennis Remington was the token opponent.

One of the scientists didn't mention weight control, while the other one spent most of her allotted time discussing the subject of weight management. She introduced three experimental studies, two of which showed a weight gain with the use of artificial sweeteners! The first of these studies involved three groups of rats: one given artificial sweeteners, one given sugars, and one given excessive fats. She showed that although the artificial sweetener caused a weight gain, it did not cause as much gain as that produced by fats and sugars.

The second study she quoted was an American Cancer Society survey of more than 78,000 Americans. This study apparently was done in an attempt to identify risk factors for cancer. Patients were asked in a questionnaire whether or not they used artificial sweeteners, and were asked to identify whether they had gained weight during the previous year. In each weight category, from the very thin to the very overweight, those who used artificial sweeteners had a greater weight gain than those who did not use artificial sweeteners.[46] She admitted that the study was statistically significant, but suggested that the amount of weight

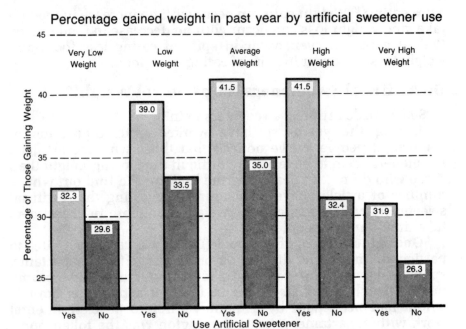

Percentage gained weight in past year by artificial sweetener use

Figure 4. Results of survey including 78,694 white women in the USA between ages 50 to 69. 21.6 percent were long term users of artificial sweeteners, 78.4 percent were non-users. Note that in each weight category, artificial sweetener users had higher incidence of weight gain than non-users in the year prior to the survey. Drawn from data presented by the American Cancer Society.[46]

gain really was not that high, since the average person only gained a few pounds per year from using artificial sweeteners. (See Fig. 4.)

The third study was presented as evidence that aspartame could be useful in helping people reduce the total number of calories they eat. Instead of quoting her study, we would like to describe one that was designed the same, that had the same results, and has been reported in the medical literature.[47] This study will give you some appreciation of the type of research done by proponents of artificial sweeteners to make these products appear useful.

Six men, with an average age of forty and who were described as having "limited economic resources," were confined in a metabolic unit of a hospital and given free access to a wide variety of highly palatable foods twenty-four hours a day. They even had a bedside refrigerator well stocked with goodies. The diet contained "significantly more meat dishes, desserts, and sweetened beverages than subjects reported eating at home." During the six days of this "baseline diet," these men ate an average of 3632 calories daily, and gained an average of 1 3/4 pounds.

The study design included many features well known to cause maximum caloric intake and weight gain. The subjects were poor, and would be expected to eat plenty in response to the free food. They were older than the usual college students normally used in studies of this type, and would be expected to have a lower metabolic rate, lower energy requirements, and gain weight more easily than younger men. They were confined in a small space, like farm animals penned to decrease activity and promote weight gain. They had a wide variety of food that was high in sugar, high in fat, highly palatable, and always available. As shown earlier, these conditions can cause increased weight even when the amount of food is limited.

The average daily energy requirement for inactive men of this age would be fewer than 2500 calories daily. The 3632 calories daily was like a force-feeding situation, causing some early weight gain. (Remember the force feeding study quoted earlier? After force feeding to gain weight, the subjects had to continue eating at least 2400 calories per day more than their usual diet in order to maintain the gained weight.) In this study, weight loss would be expected, then, if the number of calories dropped below 3632, even if the men were still eating more calories than they normally do.

An interesting thing happened. During the twelve days when aspartame was substituted for the sugar in some of the food and drink items, these men ingested a daily average of 3010 calories. In the first three days, the average weight dropped back below the baseline—but then the weight actually increased for the remaining nine days that the men were eating foods sweetened with aspartame, even though they were eating 662 fewer calories each day than on the baseline diet!

What was the conclusion derived from this study? Researchers concluded that "low-calorie food analogues could be useful adjuncts in weight control!" Using artificial sweeteners to promote the intake of more than 3,000 calories daily in a group of middle-aged, inactive men could hardly be considered a triumph of modern diet technology!

The bottom line is this: These men using artificial sweeteners ingested considerably more calories than they did under their normal circumstances. If they had started with the diet containing aspartame they would definitely have gained weight. The force feeding of the high fat, high sugar diet before the study is the only thing that prevented weight gain during aspartame use.

Of further interest is the fact that this exact study with the same diagram was included in a book produced by the manufacturers of aspartame.[48] In this book, the weight line was eliminated from the diagram (See Fig. 5), in an apparent attempt to make aspartame appear more effective.

The issue of whether or not artificial sweeteners are of any value in weight management has been evaluated by a number of different authorities, all with the same results: "There is no evidence to show that artificial sweeteners are useful in weight reduction."[49,50,51,52,53,54,55,56]

The best that can be concluded about the use of artificial sweeteners is that the amount of weight gain they cause is less than what you can gain from a diet high in fats or refined sugars. This might be small consolation for those who are interested in maintaining their weight or losing weight, rather than continuing to gain.

Another interesting feature is that the weight gain caused by artificial sweeteners seems to be independent of the calories ingested. Furthermore, it may induce increased food intake. In other studies, artificial sweeteners have been thought to actually help dieters reduce caloric intake.[57,58,59,60,61] In none of these cases, however, has effective weight loss been demonstrated. To understand how this weight gain can occur in spite of reduced food intake, it is useful to understand more about the nature of obesity.

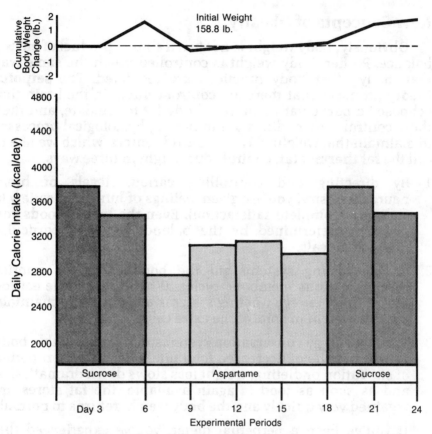

Figure 5. Mean caloric intake during experimental periods lasting three days each. During periods 1,2,7, and 8 subjects received the sucrose-sweetened diet and during periods 3 to 6 the aspartame-sweetened diet. The bar graphs in the lower section of the figure show the average daily caloric intake for the six subjects. The upper section of the figure represents changes in body weight over the course of the experiment. The dotted line represents the average initial weight of the six subjects and the solid line represents the average body weight at the end of each experimental period. Drawn from data presented by Porikos, et al.[47]

New Concepts of Obesity

Obviously, body weight is not just a simple product of energy balance. Rather, body weight is controlled much the same way that many other body functions are regulated. The **setpoint theory** proposes that there are control centers in the brain that "choose" a particular amount of body fat to regulate, and then these control centers direct a number of physiological processes to maintain that weight.[62] These control centers, which we like to call the **fat thermostat,** control your weight in three ways:

1. By directing and controlling various levels of brain neurochemistry, you are given feelings of hunger or satiety (a feeling of complete satisfaction). Even the *type* of foods you desire is determined by the balance of these particular neurochemicals.

2. Energy-wasting systems rid the body of excess calories through various metabolic cycles. When thin people eat too much, these energy-wasting systems allow them to maintain a stable weight in spite of the extra calories.

3. Various energy conservation systems maintain a stable body weight in spite of decreased food intake. After a short period of starvation or dieting, weight loss slows down dramatically, and as soon as food is again available, the fat stores are regained very quickly and the body weight returns to normal.

If you've been a perpetual dieter, you've experienced this first-hand. Each new episode of dieting results in slower weight loss than did the previous diet.[63] And, worse yet, you don't have to eat as much to maintain your weight after an episode of dieting.[64,65] A chronic dieter soon becomes well equipped to combat starvation, with a body that gives up fat very reluctantly, maintains it with a relatively low food intake, and gains any lost weight back very quickly on a small number of calories.[66] In fact, an episode of dieting often causes you to put on more fat to better protect you against some future period of low food intake.[67]

Remember the experiment with the simulated prisoners of war? After the starvation period, subjects resumed normal eating and they gained only slightly more than they weighed before the experiment, but they gained considerably more fat. (See Fig. 3.) How does that translate? What it means is this: during a dieting episode, muscle and other protein tissue is lost. You gain back fat first, and your muscle is regained only after a prolonged period of adequate food intake.[68] Why? Your body is trying to protect itself against a possible period of starvation in the future. By giving up muscle tissue, which has a high metabolic need (muscle requires much more energy to maintain than does fat), and replacing the muscle with fat tissue (which requires little energy to maintain), your body can better withstand starvation.

There are a number of mechanisms involved in this general response to starvation, two of which are particularly interesting. A special enzyme called adipose tissue lipoprotein lipase, found in fat tissue, plays a major role in fat storage. This enzyme seems to help form the fat from other substances, and to deposit these fats within the fat cells. In response to a period of low food intake, this enzyme increases to high levels, contributing to the rapid weight gain often observed after dieting.[69,70]

Another mechanism of particular interest in weight gain is the action of **insulin.** Insulin has been called the "fat hormone," and will make experimental subjects fat, even without adding any additional food. Most overweight people constantly have higher levels of insulin than do thin people, and they also have higher levels of insulin after eating. Overweight people also have a relative **resistance to the action of insulin,** which means that it requires more insulin than usual to get sugar into the cells of the body where it can be burned for fuel.

Because of the extreme importance of insulin in regulating body weight, and because of the way artificial sweeteners can affect insulin, you should understand more about the way insulin works.

After eating food which can be converted into sugar, your blood sugar gradually begins to rise. Along with the rise in blood sugar, more insulin is produced. As insulin is infused into the bloodstream, it binds to receptor sites on the outside of each cell, and helps to transport the sugar into the cells. When you eat any sweet tasting food, a surge of insulin is released from the pancreas into the bloodstream even before the blood sugar begins to rise.

This surge of insulin in response to sweet taste has been referred to as the **cephalic phase of insulin release.** The cephalic phase of insulin appears to prime the liver to store sugar in the form of glycogen.[71,72] All of the sugar absorbed from the intestines passes into the bloodstream, through the portal circulation into the liver, and then into the rest of the body. With appropriate cephalic phase of insulin release, a good share of the sugar is retained in the liver, where it is stored as glycogen. Without this appropriate cephalic phase of insulin, most of the sugar passes right through the liver into the rest of the body, causing a rapid rise in blood sugar.

There is a significant difference in insulin and blood sugar response between a normal person and an overweight person. (See Fig. 6.) In the normal person with the regular cephalic phase of insulin release, the blood sugar level rises relatively gradually, since much of the sugar remains in the liver. Because of the gradual increase of blood sugar, only a small amount of insulin is necessary to return the blood sugar to a more normal level. In the overweight person, something is wrong with the sugar metabolism. The cephalic phase may be impaired, thus very little sugar stays in the liver;[73] most of it stays in the bloodstream, causing a rapid spike in blood sugar, which in turn triggers excessive production of insulin. This combination of high insulin and high sugar seems to be the major defect that causes fat gain. Much of the sugar is converted to fat, which is then stored in the fat cells, where it is less accessible for quick energy needs.

After the food from the previous meal has been digested, and is no longer providing a continuous flow of sugar into the bloodstream, a normal person is able to break down the sugar stored in the liver to keep the blood sugar stable for many hours. The overweight person, on the other hand, with limited sugar storage in the liver, begins to get hungry, especially for sugar, as the blood sugar begins to fall.

This cephalic phase of insulin release is a critical factor in normal sugar metabolism.[74] Anything that interferes with it may change a thin person's metabolism to one like a fat person's.

So what can interfere with the cephalic phase of insulin release? A major factor is going for prolonged periods with little or no food. A number of hours after eating, the glycogen stores in the liver become very low. Muscle tissue and fat stores are broken down to provide the needed energy. When food is again ingested, the first priority appears to be to restore the depleted

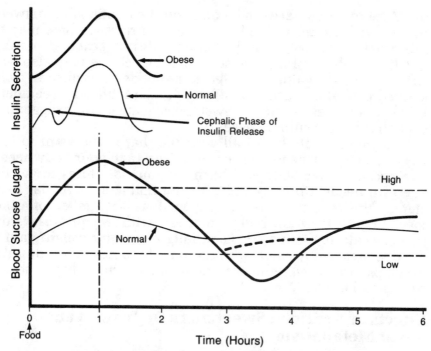

Figure 6. Insulin and blood sugar response after eating in normal weight humans compared to obese humans.

muscle and fat tissues. The cephalic phase of insulin release is not activated,[75] and the blood sugar rushes into the bloodstream to the various tissues,[76] where it quickly starts to make up the deficit created by a prolonged period without food. The rapid rise in sugar, together with the high levels of insulin it triggers, may be responsible for driving much of the sugar into the fat stores. Your body will replace the fat stores first, and will replace the muscle tissue later.

This failure to produce a normal cephalic phase of insulin in response to prolonged periods without eating may contribute to the obesity produced in rats that are fed only once a day. People who miss meals or eat only once a day may also trigger a fat gain, partly by interfering with the cephalic phase of insulin release.[77,78] Even the common practice of eating small meals or cutting calories may interfere, and cause long-term fat gain.

More research needs to be done on the relationship between obesity and the cephalic phase of insulin release; there may be other factors involved as well. Some studies suggest that an exaggeration of the cephalic phase of insulin release may also contribute to weight gain.[79,80,81] High-sugar foods may cause an exaggerated cephalic phase of insulin release. Artificial sweeteners, because of their extreme sweetness, cause an exaggerated cephalic phase insulin release.

You've probably heard all your life that if you want to lose weight, all you have to do is eat less food than your body needs, and presto!—your body will burn up your fat stores to get the energy it needs. Unfortunately, it's not that simple. As has been stated, the insulin response—as well as a number of other mechanisms in your body—interfere with weight loss from reduced-calorie dieting.[82,83] Especially at risk for weight gain is the one-meal-a-day eater, who not only becomes fatter, but also develops impaired glucose metabolism and high serum cholesterol levels.

Effects of Artificial Sweeteners On Insulin and Sugar Metabolism

Now for the role of artificial sweeteners:

Experimental rats who are given saccharin to drink will trigger a cephalic phase of insulin release, and will have a rapid drop in blood sugar.[84,85,86,87,88] Various compensatory mechanisms within their bodies will eventually get the blood sugar back to normal. Humans also have a drop in blood sugar after having artificial sweeteners.[89,90,91,92] This response occurs when artificial sweeteners are given by themselves with no additional calories, and is, at least in part, a **conditioned response.**[93,94]

A good example of a conditioned response is a study done by Pavlov, who trained a group of dogs by bringing them food while he was ringing a bell. Soon the dogs were conditioned so that the ringing of the bell by itself would cause them to salivate in anticipation of eating. If Pavlov continued to ring the bell but stopped bringing the food, the dogs soon stopped salivating when they heard the bell. In the same manner, rats continually given artificial sweeteners without food will soon stop producing insulin in response, and the cephalic phase of insulin release is gradually extinguished.[95,96] This may then interfere with sugar metabolism and contribute to weight gain.

If you regularly eat highly sweetened foods or drinks, your body becomes accustomed to having very sweet things to trigger the cephalic phase of insulin release. If you then eat food that is not as sweet, even though there may be plenty of carbohydrate present in those foods, the cephalic phase of insulin may not be triggered adequately.[97,98] That's why artificial sweeteners may pose a particular hazard. Since artificial sweeteners are many times sweeter than table sugar, they may be even more likely than sugars to abolish the cephalic phase of insulin release in response to foods which are less sweet. These unsweetened foods may therefore produce higher blood sugar levels and higher insulin levels than they normally would, contributing to weight gain.

Possible Mechanisms For Weight Gain Associated With Artificial Sweetener Use

Artificial sweeteners may contribute to weight gain in any of the following ways:

1. Artificially sweetened products with few or no calories will trigger an inappropriate cephalic phase of insulin release. The blood sugar tends to fall, which then triggers the release of various compensatory hormones, which keep the blood sugar from falling too low. These hormones, including adrenalin, norepinephrine, and cortisol, are referred to as "insulin counter-regulatory hormones," and they tend to block the action of insulin and counteract its tendency to lower blood sugar. In fact, they tend to raise the blood sugar. In their presence, more insulin is needed to balance the blood sugar. These extra counter-regulatory hormones contribute to insulin resistance, which may then cause weight gain.

2. By routinely using artificially sweetened products with no calories, especially diet drinks, the cephalic phase of insulin may eventually be extinguished, and your body's sugar metabolism may function like that of a fat person's, with high blood sugar, excess insulin production, and excessive fat deposits.

3. Artificially sweetened food is sweeter and more palatable, and may cause an exaggeration in the cephalic phase of insulin release. This seems to cause increased weight.

4. Artificial sweeteners may help you to eat less food, or eat less often, if you use them along with a consistent, determined effort to resist eating in response to hunger. Lowered food intake may cause some temporary weight loss, but may also trigger a number of starvation defenses, which may lower the metabolic rate, raise the setpoint, and promote weight *gain* in the long term, not weight loss. The reduced food intake, especially missed meals, may interfere with cephalic phase of insulin release, further encouraging weight gain.

5. Artificial sweeteners, especially aspartame, may alter the level of brain neurochemicals. These neurochemicals play a profound role in regulation of eating, hunger, desire for specific foods, and in regulation of fat stores.[99,100,101] This will be discussed in detail in the next chapter.

6. The continued use of artificial sweeteners has been shown to cause a "chronic, gradual increase in EOP levels" (opioids or morphine-like chemicals produced within the brain).[102,103] Opioids are well known to cause an increase in sweet consumption,[104] fat consumption,[105,106] and total food consumption.[107] Obese people are known to have higher levels of opioids in their blood and spinal fluid.[108,109] Even with no extra calories, a higher ratio of fats and sugars are well known to trigger weight gain.

7. Artificial sweeteners may cause an addiction-like condition. A few hours after an addictive substance is ingested, a withdrawal process begins. As the withdrawal symptoms get worse, there is an increasing desire to consume that substance again. In the case of artificial sweeteners, the addictive process may be more for sweets than for the actual product itself, and the ingestion of any kind of sweets may stop the unpleasant withdrawal symptoms. This may lead to extra sugar intake, but also contributes to weight gain in another way. During withdrawal, there appears to be a number of stressor-type hormones released that cause symptoms like restlessness, agitation, irritability, and anxiety.

These stress hormones are also insulin counter-regulatory hormones, and may cause relative insulin resistance, which could contribute to weight gain.

The conclusion?

There's absolutely no evidence that artificial sweeteners can help you lose weight—even on a short-term basis.

But that's not all. There is *plenty* of evidence that artificial sweeteners can cause—or at least contribute to— weight *gain!*

But don't let that discourage you if you have relied on artificial sweeteners. Now we know exciting new things about weight management—and that knowledge has enabled us to come up with some new ways of managing weight problems. Best of all, you can lose weight comfortably while still eating plenty of satisfying food.

Want to know more about the possibilities? Keep reading!

CHAPTER TWO

DO ARTIFICIAL SWEETENERS SATISFY YOUR DESIRE FOR SWEETS AND HELP YOU DECREASE YOUR SUGAR CONSUMPTION?

If you're like most people, you know that eating too much sugar is unhealthy.

But if you're like many people, you have a sweet tooth—and you want to satisfy it.

The logical answer?

Choose something sweetened with an artificial sweetener. That would replace some of the sugar in your diet, and would be a healthy alternative—right?

Not necessarily—for a number of reasons.

Do artificial sweeteners help you eat less sugar?

There isn't a simple answer to that question.

Since more than 150 million Americans have been estimated to use artificial sweeteners,[1] we should be able to study sugar intake and, as a result, figure out whether artificial sweeteners have really had an impact. But let's take a look at what we know.

According to the United States Department of Agriculture, from 1975 to 1984 our annual consumption of artificial sweeteners increased from the sweetness equivalent of 6.1 pounds of sugar to 15.8 pounds of sugar per person. If artificial sweeteners really work as a sugar substitute, sugar consumption should have gone down during that same time period. Not so! According to one source, annual sugar consumption went *up* from 118.1 pounds per person to 126.8 pounds.[2]

We have new weight control patients keep an accurate record of what they eat for one week before they start any specific weight management program. We ask them to eat as they normally would when not dieting, and they note their degree of hunger, their mood, and their behavior at the time of eating. Our main goal is to help patients identify problem areas and change defeating behaviors, but we found something interesting as a bonus: many of these overweight people were drinking huge quantities of diet drinks.

We were astounded. Some of these "dieters" drank twelve to twenty-four cans of diet soda *every day.* One patient was drinking thirty-six cans of diet pop daily, and they all contained caffeine! It didn't take a genius to identify the source of his anxiety and sleep disturbance.

You'd expect that someone drinking that much sweet soda would eat few sweet foods, right? Surprisingly, that's not the case. The patients who drank the most diet pop usually ate the most foods highly sweetened with sugar, too. Usual favorites were regular pop (even though they were getting plenty of diet pop), cookies, ice cream, cake, donuts, other pastries, and candy bars.

Although these patients were eating high-sugar, high-calorie foods, some were eating relatively few total calories. Why? They somehow succeeded in denying their natural hunger and restricting themselves to a low-calorie regimen most of the time. When they occasionally failed, they would binge on high-fat, high-sugar foods. Sadly, that combination—food restriction alternating with indiscretion—is the perfect setup for weight gain.

On our history forms routinely filled out by all new patients, we specifically ask whether they drink soda pop and diet drinks. New patients are asked specific questions about their use of artificial sweeteners. Many who did not like the aftertaste of saccharin now find that diet drinks sweetened with aspartame taste just like regular soda—and they have started to drink huge amounts.

What have we found out about these patients? Many admit to a definite weight gain since they started drinking diet drinks. But are they eating less sugar? No. They often admit—with some surprise—that they are using more sugar than ever before.

A few studies have shown that you can substitute products sweetened with artificial sweeteners for those containing sugar, at least for a short term under careful laboratory-controlled circumstances. This just doesn't seem to work in the real world, though. Most evidence suggests that those using artificial sweeteners usually increase their intake of sugars. We'll discuss the reasons why later.

Do artificial sweeteners satisfy your desire for sweets?

In a land of plenty, it is easy to overlook the fact that our most important need is for food. For many of us, eating is so automatic that we hardly think of food. Those who frequently diet, however, often describe thinking of little besides food. Patients have jokingly stated that after three days of dieting, they would gladly kill for a piece of chocolate cake. That's not as extreme as it sounds—people deprived of food have been known to commit every imaginable crime, including murder, to get food.

Contrast this extreme hunger situation to the feeling you get after a typical Thanksgiving Day dinner. Although your stomach may be somewhat uncomfortable from eating too much, and you may suffer a little guilt, you generally feel great. You feel completely satisfied, calm, relaxed, peaceful, happy, and contented. Even if the kids raise a ruckus you don't really care, because you feel great. You feel so relaxed, in fact, you often go to sleep.

The key? Satiation, or satiety—the state of having your appetite completely satisfied. It all begins in your brain, which sends out the messages your body gets when you feel hungry.

When you eat, you taste, smell, chew, and swallow the food. Sensors signal your brain that these activities have taken place. Your stomach stretches (distends) a little from the food you have eaten, and your brain gets another message to help signal satiety. But having a full stomach alone doesn't do it—if food was pumped into your stomach through a tube, it would take considerably more food to produce a sensation of satiety than if you had chewed, tasted, and swallowed those same foods.

From the stomach, food enters the intestines, and various intestinal hormones and digestive enzymes are released, which also influence satiety. The balance of protein, fat, and carbohydrate is somehow measured, as well as the total food energy (calories). This information influences satiety.

From the intestines, digested food enters the bloodstream; from there it is either utilized by your body, stored as sugar stores (glycogen) in the liver or muscles, or stored as fat in the fat cells. Various blood nutrient levels and various stores of energy are also measured, and your brain uses this information to determine how hungry or satisfied you are from moment to moment.

How hungry or satisfied you are—and even which foods you desire—seem to be controlled by chemicals in your brain, and their effects are both important and fascinating.

Neurohormonal involvement in eating.

Beta-endorphins (for simplicity, we'll call them endorphins) are chemicals produced in your brain that have pain-killing abilities similar to morphine. Endorphins are involved with the sensation of satiety. The contented, relaxed, happy, feeling you experience after eating a full meal is very much like the feeling described by morphine users.

The sweet taste of food is at least part of what triggers production of endorphins. Saccharin has been clearly shown to cause endorphin release in rats. When rats are given saccharin in their drinking water, they drink considerably more water than they need. They even start to respond to their drinking water like a drug abuser does to drugs—and they drink the water at abuse levels. The rats work very hard to get the sweetened water, so obviously they get some kind of pleasure from it. What helps the rats break their "addiction?" Naloxone, a drug that blocks the action of both morphine and endorphins, does the trick. Why? Because with naloxone, the rat's endorphins no longer cause any drug-like effect.

Besides contributing to the comfortable feeling after eating, endorphins play other roles in relation to weight regulation and eating. Scientists have now discovered at least three categories of opiates (or opioids) and many different types of receptor sites for these opioids.[3,4] The same opioid may cause opposite effects in different doses. One dose may stimulate eating, and another may stop it. The increased levels of endorphin caused when you eat saccharin, for example, triggers an increased desire for and an increased intake of sweet solutions.[5]

Obese people have been shown to have lower levels of one type of endorphin, but higher levels of other endorphins, in their blood and spinal fluid.[6] Since obese people are often restrained eaters (they miss meals and force themselves to go hungry)[7], their reward and satiety endorphins may be low. They may have high levels of the endorphins that trigger insulin production,[8,9] which may be an important factor in their obesity.

Endorphins seem to cause an addiction-like state. The good feeling that runners get after a prolonged period of running—the famed "runner's high"—seems to be caused by endorphins. These endorphins cause some runners to feel so good that they run in spite of painful injuries just to get the reward of the good feeling. Apparently, some people use sweets in the same way— to get the good feeling produced by endorphins in response to sweet taste. If you scoff at the idea of an "addiction," consider this: people who use artificial sweeteners generally use them several times a day, often in huge quantities—all typical features of an addiction-like state.

Whether artificial sweeteners can be called addictive is debatable, since they don't seem to cause severe withdrawal symptoms. Users who try to stop certainly have a desire and cravings for them, but don't seem to have physical withdrawal symptoms as severe as those characteristic of other well-known addictive substances like drugs, nicotine, alcohol, and caffeine. To avoid any conflict, we will refer to the effects of artificial sweeteners as "addiction-like" effects.

One more point needs to be made. The addiction-like response to artificial sweeteners results from the sweet taste, and sweeteners have to be perceived as sweet before this response occurs. The latest rage in sweeteners, aspartame, does not appear to taste sweet to rats. They show no preference for it, they do not abuse it, and it does not appear to give them pleasure or cause any endorphin response.[10,11] Even those experimenters who are paid by the aspartame manufacturers admit that rats do not perceive aspartame as sweet. That raises plenty of ethical questions about aspartame research: if the experimental animals don't perceive an artificial sweetener as sweet, are they reasonable substitutes for humans in research?

Another brain chemical with a significant impact on eating is serotonin. After you eat a meal that contains adequate amounts of carbohydrate, the level of serotonin in your brain goes up—and the serotonin is part of what helps you feel completely satisfied, happy, relaxed, and content after eating. A few hours after you eat, the serotonin level falls, and you start to feel hungry. Not only do you get progressively more hungry as serotonin levels fall, but you specifically crave foods containing carbohydrates.

Serotonin, then, seems responsible for regulating the amount of carbohydrate you eat in relation to other foods. In nature, carbohydrate is seldom found all by itself (one exception is honey, which is refined by bees from plant materials). While most plant material that we use for food is mostly carbohydrate, it also has protein and a small amount of fat. With the exception of liver, flesh foods and eggs contain protein and fats, but virtually no carbohydrate. What does that tell you? When you eat, your serotonin level won't get enough of a boost unless your meal contains enough carbohydrate. In other words, if your meal contains only protein and fats, you will still feel unfulfilled or discontent until you eat some carbohydrate. (There is an exception: if you continue to eat very little carbohydrate on a long-term basis, your body produces ketones, which artificially suppress hunger and the desire for carbohydrates.)

Although serotonin is influenced most by the carbohydrate in your diet, it is made from an amino acid, tryptophan, which comes from protein. The ratio of amino acids in your brain has a great deal of influence on your hunger, specific food desires, and your mood and behavior.[12]

After a meal containing carbohydrate, insulin is released into the bloodstream. In addition to its role in carbohydrate metabolism, insulin also affects amino acids. In response to increased insulin, the amino acids leave the bloodstream, and are taken into various cells. Insulin affects competing amino acids more than it does tryptophan, so the result is a higher ratio of tryptophan in the bloodstream—and, therefore, in the brain. The result? More serotonin is produced. As serotonin levels rise within the brain, you begin to feel satisfied, slow down your eating, and eventually stop eating entirely. Hours later, after the food you have eaten is digested, the insulin level drops, the amino acid ratio in the blood changes, serotonin level begins to drop, and the desire to eat something containing carbohydrates gets strong. How do you respond? You eat.

Obviously, serotonin is important in the beautifully orchestrated eating response directed by the brain. When you're eating healthy food, the cycle continues, uninterrupted, day after day: you get hungry, you eat, you feel content and satisfied for hours, and then you get hungry again.

That's the way nature intended it.

But that's *not* what happens when you use artificial sweeteners.

Here's how artificial sweeteners can play havoc with your natural hunger cycle. If you are hungry and you drink a diet drink, the sweet taste triggers a release of insulin in anticipation of carbohydrates being ingested—even though the diet drink contains few or no calories. The insulin release can cause a chain of events that boost brain serotonin levels. The sweet taste may also trigger the release of endorphins. The tasting, swallowing, and filling up of the stomach may also decrease your hunger.

What happens, then?

The diet drink may give you a certain degree of satisfaction and take your mind off eating—temporarily. Because the insulin surge after sweet taste is small and short-lived compared to the amount released after you eat regular food, the rise of serotonin in your brain is also minimal. Besides that, the insulin tends to drop the blood sugar a short time later, which signals you to eat. Because none of the many other elements of satiety are satisfied, hunger gradually intensifies—and, eventually, you will need to eat.

When the diet drink is sweetened with aspartame, something else happens. Aspartame is a combination of two amino acids, which are broken down as it is digested. One of these amino acids, **phenylalanine**, competes with tryptophan, causing a conflict. The result? Your brain doesn't produce as much serotonin. So diet drinks sweetened with aspartame may not work as well as other diet drinks to temporarily relieve hunger. But because most people think that aspartame tastes better, they probably enjoy better endorphin response.

Either way, one thing's obvious: the hunger relief you get from guzzling a diet drink is temporary at best.

Several other brain hormones also have an influence on body fat stores. Various parts of the brain that control body weight use **dopamine, norepinephrine, and epinephrine (adrenalin)** as neurotransmitters.[13,14] These same hormones also influence heat production, which can eliminate extra calories and prevent weight gain from overeating.[15,16] In overweight people, these heat-producing cycles are impaired.[17,18] Some of these hormones may also affect the degree of insulin resistance.

Overweight people have been shown to have lower than normal serum concentrations of the amino acid tyrosine, which is used by the brain to produce dopamine, norepinephrine, and adrenalin.[19] Anything that causes a depletion of some of these brain chemicals can cause obesity.[20] Since aspartame also tends to lower blood levels of tryptophan, it may cause lower levels of these important brain hormones,[21] and may contribute to weight gain in a number of different ways.

All of this insight about brain chemicals and artificial sweeteners sheds some light on the weight issue. But can artificial sweeteners also make you feel more hungry? Studies indicate they can!

Artificial sweetener effects on hunger.

A number of studies have been done in an attempt to evaluate the effect of artificial sweeteners on appetite, hunger, food intake, and so on, even though some of these effects are difficult to measure scientifically. Dr. Blundell in 1986[22] described a study in which human volunteers under various circumstances were given glucose, water, or aspartame. Researchers then evaluated how hungry the volunteers were in a series of assessments. One of these assessments measured the degree of pleasure a volunteer experienced drinking a sucrose (sugar) solution after first having had a solution of either water, glucose, or aspartame.

Have you ever eaten two desserts, one right after the other? If your experience was typical, you probably didn't enjoy the second one nearly as much as you did the first. Why? Because the hungrier you are, the more pleasant something sweet tastes to you. It was like that with the research volunteers—as expected, the sugar solution didn't taste as good after first drinking another type of sugar (glucose). Those who drank water first noticed little effect—the sugar solution tasted very sweet. Drinking an aspartame solution first caused a reaction right in the middle: the second drink didn't taste quite as sweet as it did after drinking water, but it tasted sweeter than it did after drinking a glucose solution.

The obvious conclusion? Aspartame helps reduce hunger—a little.

But Dr. Blundell did not stop there: he measured hunger in three other ways. He assessed the motivation to eat and the sense of fullness achieved after a drink of water, glucose solution, or aspartame. No one was surprised by what happened after volunteers drank water: there was virtually no change. No one was surprised, either, by what happened when the volunteers drank the glucose solution: volunteers felt less motivation to eat and felt more full. The big surprise came with the aspartame solution: volunteers felt a greater motivation to eat and reported feeling less fullness!

The result? An apparent conflict: one component of hunger seems to be decreased by aspartame, and other components of hunger seem to be increased by aspartame. The most meaningful result of all this research is behavior: does aspartame cause people to eat more? According to Dr. Blundell, yes. Specifically, it causes people to eat more foods that contain sugar.

In another study[23], Dr. Sclafani showed that rats given access to a type of sugar solution called polysaccharide became fatter than those eating only rat chow. This sugar solution is rather bland-tasting compared to sucrose. Making it sweeter by adding saccharin to it caused rats to eat more of it, increasing their total calories.

If you don't have a weight problem, you probably aren't that concerned about the fact that artificial sweeteners cause you to eat more. You probably use them because, simply, foods sweetened with them taste good and bring you pleasure.

But do they really bring you pleasure?

You might be surprised by the answer.

Do artificial sweeteners give pleasure?

Obviously, the mere acts of eating and drinking bring pleasure. A hungry baby cries and screams until he is fed—and then he quiets down, relaxes, smiles and laughs, or goes to sleep. A very hungry child acts in much the same way: he whines, complains about feeling hungry, announces that he is tired, and begs for something to eat. You can distract him by something other than food only briefly—but he soon becomes totally preoccupied again with the need for food. Once fed, he is like a different child—happy, relaxed, easy to please, and reasonable.

As an adult, you respond to hunger the same way a child does, but you've learned to control yourself and to be nice to the people around you, even when you're feeling miserable. You can generally distract yourself better by keeping busy, and you can ignore the hunger drive better, so you probably aren't as in touch with your feelings and needs as a child is. Those who miss meals, frequently reduce their food intake, or eat small amounts when they do eat, are usually even more out of touch with their basic needs. They deny and suppress their hunger so often that they don't realize that their discomfort stems from a lack of food.

When you are uncomfortable due to a need for food, eating can produce a great deal of pleasure. The smell and taste of the food produces pleasure, the sense of fullness in your stomach feels good, and many changes in your brain and body in response to eating give you pleasure. If you eat until you are completely satisfied, you'll experience even more pleasure—along with a feeling of well-being and contentment that can be attained in no other way.[24] These good feelings can linger for hours after you eat.

If you are hungry and you drink a diet drink, you'll experience a sense of pleasure. The diet drink might even take the edge off your hunger, and allow you to forget about eating for a short time. But the diet drink will only give partial relief of the hunger, and will last only a short time. Soon the biological need for food again signals to you—and it will be even stronger, because your need for food becomes progressively greater. But that's not the end of it: the artificial sweetener may cause a drop in your blood sugar level, and you'll feel even more weak and hungry.

The small and fleeting pleasure you get from a diet drink when you are hungry is a very poor substitute for eating real food, which provides greater pleasure and a longer lasting sense of satisfaction. If you continue to use artificially sweetened products as a substitute for real food, you are likely to gradually develop a wide range of very unpleasant symptoms associated with restricted food intake. A vicious cycle results. You feel gradually worse, but the diet drink temporarily relieves your symptoms, making you think that they are really pleasant and fun to drink. So you keep using them to keep yourself from feeling so miserable.

If you continue to restrict your food intake, you will eventually find that even a large meal will no longer completely satisfy you or give you pleasure for very long. In the conscientious objector study mentioned in the first chapter, a great number of unpleasant symptoms were reported from several months of reduced caloric intake. When these men were finally allowed to eat all they wanted, they couldn't get a feeling of satiety even with large quantities of food. They wanted more food and felt restless and deprived. It took several months of eating a great deal of food before they began to again feel a complete sense of relief and contentment after a full meal. Many people who persistently restrict their food intake can no longer even remember what it is like to feel complete and utter contentment from a meal!

Another problem arises from using highly sweetened products of any type, whether used by themselves with few calories (as in a diet drink), or whether used to sweeten real food. When you frequently eat or drink highly sweetened products, your senses become used to the extremely sweet taste. The sweetness causes a number of changes to occur—including release of insulin and release of endorphins—and contributes to the sense of satiety caused by eating. Foods that are not as sweet may no longer cause adequate insulin and endorphin release, and may no longer be satisfying to you. That's not all: foods that are not as sweet may no longer taste good. Using highly sweetened foods on a frequent basis will thus rob you of the pleasure normally derived from eating good wholesome food!

And what about weight gain?

As we pointed out in the first chapter, there seems to be a clear relationship between weight gain and the use of artificial sweeteners. In spite of this, promoters of diet drink products try to portray a very different picture in their advertising. They show very thin, extremely attractive young people using diet drinks, having fun with each other, pleasure and happiness radiating from their faces. The message? Diet drinks will help you become thin, attractive, and desirable—and will deliver pleasure and fun as part of the package!

A more accurate picture?

A fat person sits at home alone, watching other people have fun on television, wallowing in self-pity in a sea of empty diet drink cans. The diet drinks staved off hunger for awhile, but the empty chocolate chip cookie bag and crumb-spattered face vividly tell the story of failure. Instead of having fun, like the people in the diet drink commercial, she has been groping in the bottom of the bag for more cookie crumbs. Instead of joy, bitter tears of grief well up in her eyes. Instead of a face radiating happiness, her facial expression shows self-loathing, disgust, guilt, and total frustration. She seems to be thinking, "How can I fail so miserably when I so desperately want to be thin and when I try so hard? What's wrong with me?"

If you've never struggled with your weight, you probably think we're exaggerating. But if you have fought a weight problem, you know that scenarios like this are all too real and all too common.

Not everyone who brings a diet drink to his lips will inevitably become fat and miserable—but if you choose to use artificially sweetened foods and drinks, you owe it to yourself to learn the inevitable risks and problems.

You've learned a few in these two chapters—but we've just scratched the surface. There's more—and to find out what, turn the page.

CHAPTER THREE

DO ARTIFICIAL SWEETENERS HELP YOU EAT A HEALTHIER DIET?

Today, we seem to live in a utopia as far as food is concerned. Why?

Healthy food is available year round from all parts of the world. Supermarkets and convenience stores are open around the clock. Fast food outlets and food retailers seem peppered every few blocks in every town and city across the nation. Food is relatively inexpensive. Virtually every home has a stove and a refrigerator with a freezer, and many have microwaves that can zap up a snack within seconds. Simply stated, you can choose whatever you desire—hot food, cold food, or any combination!

With all that delicious and nutritious food at our fingertips, it should be very easy for each of us—even the poorest among us—to eat a healthy diet.

So how does the average American eat?

The average American adult eats 2,000 calories a day. Of that, more than 40 percent (more than 800 calories) comes from fats. More than 600 calories daily come from sugar. Since fats and sugar have almost no nutrients other than calories, we have to get all the vitamins, minerals, and other nutrients essential for optimal health from fewer than 600 calories a day. That would be bad enough—but for many, those remaining calories come from refined wheat flour (that has had many of the nutrients removed) or from flesh protein, which provides no fiber and few of the other essential minerals or vitamins.

Does the way we are eating have any influence on our health?

Absolutely!

You've seen the way the average American eats. Add to that a relatively inactive lifestyle. The result? Most will become fatter. In fact, that's already happening to the average American. Most alarming is the rapidly increasing rate of obesity among children. In the last 15 to 20 years, the incidence of overweight among children has risen a staggering 54 percent!

Since most people mistakenly believe that weight gain is the result of too many calories, they wage war by cutting their own calories and feeding their children less. The result? Their already shaky nutritional status is further threatened—and, even worse, they cling to their extra pounds.

Several important pioneers in medical research have shown that there's more to the relationship between nutrition, weight, and health. One of these, Nathan Pritikin, developed a heart problem while in his early forties. He had high cholesterol levels, high blood pressure, and significant blockage of his coronary arteries kept adequate blood from getting to his heart muscles. He wanted answers, but his doctors told him there was really nothing he could do. His prescription? To rest more often and avoid exertion.

It was the late 1950's, and Pritikin went to the U.C.L.A. nutrition department for dietary recommendations that would help him lower his cholesterol. They told him nothing could be done. If today's technology had been available, they probably would have prescribed coronary artery by-pass surgery to improve the heart circulation. That might have helped for a few years—until the process that caused the artery occlusion in the first place again narrowed the by-passed artery or some other artery.

Pritikin was not content to lie down and die. Realizing that the medical community held out no help, he began some research on his own. He found that during World War II, when animal fats, protein, and sugar were rationed in some countries, the incidence of heart attacks was dramatically decreased, in spite of the extra stress associated with the war. He also studied eating and health patterns in many different countries, and discovered that in many underdeveloped countries hardening of the arteries was virtually nonexistent. He studied the diet of the groups that had little or no heart disease, and decided that he would eat like these groups did in an attempt to put a halt to his own heart disease.

Pritikin launched his new eating program, and the results were encouraging. His cholesterol dropped from around 300 to just over 100, and he gradually improved. He cautiously started a walking program, then began running. Within six years, his stress cardiograms were normal, and all evidence of heart disease was gone.

Pritikin committed suicide in February, 1985 rather than continue a prolonged course of treatment for leukemia. On autopsy, the pathologist found that his "arteries were amazingly clean for a man of his age."[1] Pritikin's is a remarkably clear case history of reversal of heart disease that had been diagnosed as irreversible and progressive.

Pritikin became so enthusiastic about his own results that he established a live-in health center where others could come and learn to change their own lifestyles. He also wrote a number of books. There are now thousands of case histories as dramatic as his own in which diet and exercise improved serious health problems. Not only have heart disease, high blood pressure, and cholesterol problems been cleared, but there have been dramatic improvements in diabetes, arthritis, and intestinal problems.

English physician Dennis Burkitt studied the diets and health of native Africans. He noted that they were free of most of our western diseases, and concluded that it was the fiber in their very basic, unprocessed diet—which consists mainly of grains and vegetables—that keeps them free from disease.

Why?

Burkitt believes that with a high-fiber diet, intestinal bulk from undigested food is much greater, and that it remained in the intestine for a shorter period of time compared to western diets. He theorized that high-fiber, high-bulk diets were the reason that these Africans had virtually no constipation, no intestinal cancer, and were completely free of other gastrointestinal problems such as gall bladder disease, appendicitis, and diverticulitis. He even theorized that these diets may help prevent other types of cancer, and may have other far-reaching beneficial effects, including prevention of varicose veins, heart disease, and diabetes.

Dr. Cleave, another British physician, also worked with native Africans. He did population studies in a number of groups of people throughout the world, and believed that the refining of foods—and especially the intake of refined sugars and refined fats—was responsible for a great number of our western diseases. He pointed out that the incidence of various diseases increased dramatically in some countries when they began using modern milling techniques for their wheat flour instead of using traditional methods that retained the fiber, germ, and other components discarded with milling. He pointed out that Eskimos, who traditionally had very little diabetes, developed a very high rate of diabetes twenty to thirty years after high levels of sugar and white flour became a regular part of their diets.[2]

Physician Ansel Keys was very interested in the effects of starvation on human health. In addition to the Minnesota study on the conscientious objectors mentioned in Chapter One, he also did a very extensive literature search on various groups of people who underwent starvation, either in concentration camps, during famines, during periods of food rationing, or during other incidents. He found that restricting the daily diet to fewer than 2,000 calories for men (corresponding to about 1,600 calories for women) consistently produced fatigue, weakness, poor endurance, depression, irritability, insomnia, hunger, preoccupation with food, inability to stay warm, trouble concentrating, digestive disturbances, diarrhea, frequent urination, and swelling of body tissues. He also noted a high incidence of rapid weight gain and high blood pressure after periods of food deprivation.

The work of these four individuals and many other scientists show a very clear relationship between diet and many of our health problems. It should be easy to design experimental studies to further prove the observations of these scientists and to make healthy eating recommendations based on these and other findings. Unfortunately, things are not that simple.

Large food manufacturing companies and food producer groups compete for billions of dollars spent by the food-buying public. Any changes in the population's eating habits could have disastrous consequences for these food providers. Each large company and producer group employs full-time scientists and pays millions of dollars to various university-based researchers to protect their own interests.

Any scientific evidence that a particular product or category of food causes health problems is immediately attacked by these scientists. They find fault with the studies, do some of their own studies that do not support the findings of the original studies, and in other ways attempt to resist any change in public buying patterns. A good example of this is occurring in the tobacco industry. Although it is now very clear that cigarette smoking is an absolute health disaster, tobacco companies are spending millions of advertising dollars to throw up a "smoke screen" and cloud the issues. One company starts off a friendly message to consumers with the words "Let's talk," and then proceeds to suggest that the scientific evidence implicating cigarette smoke in health problems is faulty. They sum the message up with a claim that their products are really not that bad.

These delay tactics work, and it has often taken many years before any reasonable doubt has been removed regarding the relationship of certain types of food to certain disease processes.

The government has the task of protecting the consumer, but also needs to protect various producer groups. No government agency will issue eating recommendations that will adversely affect any group of producers or manufacturers, unless there is overwhelming scientific evidence that these products are harmful. Food manufacturers and producers also have a great deal of power and influence in government, and have very powerful consumer lobby groups that protect their interest.

A good example of these influences is seen in the government policy regarding dietary fats. It has long been known that cholesterol and other fats in the diet play a major role in heart disease. It has also been determined for some time now that dietary fats contribute to cancer. Because any recommendation for dramatic reduction in fats would adversely affect the egg producers, the dairy industry, and the beef producers, recommendations have been limited to only modest reduction in the amount of eggs, dairy products, and beef we eat. Both the American Heart Association and the American Cancer Society have made recommendations that we should reduce our level of fats in the diet, but many scientists believe that these recommendations do not go nearly far enough.

One of the authors (DWR), once told Nathan Pritikin that it looked like his work was finally having some impact, since the American Heart Association had just recommended that Americans lower their fat consumption to reduce the risk of heart disease. He emphatically stated that "the American Heart Association diet does not prevent heart disease, it causes it!" Pritikin was probably right—those people vulnerable to heart disease will still be subject to heart attacks even with the new recommendations, but may live a few years longer than they would have with a diet even higher in fats.

A similar situation exists relating to dietary influence on cancer. There is a tremendous difference in cancer rates from one country to another. In 1984, the country with the most cancer reported had ten times as much cancer as the country reporting the least. In breast cancer, some countries (England, Wales, and Denmark) reported more than 100 times as much as other countries (Honduras and Nicaragua).[3] It is generally agreed that most of these differences relate to differences in lifestyle, and that diet is important. There is some good evidence from population studies as well as animal experimentation that fats in the diet contribute to cancer, and that some vitamins and other food components may be preventative. In spite of all this research—conducted during the past fifty years—the conclusion of a recent workshop on "Nutrition in cancer causation and prevention" was that ". . . there remain many disagreements on the role of individual components, so that in many cases conclusions may be premature . . . . it is not possible to supply any definitive recommendations."[4]

In spite of the lack of scientific agreement on the relationship between diet and good health, a number of official dietary recommendations have been made for the American public by various government and health agencies.[5,6,7,8,9,10,11,12] These recommendations are in almost complete agreement. Official recommendations from other countries are very similar.[13,14,15,16,17,18] Although the direction is clear, the amount of change to make for optimal health may not be specified, or may not go far enough. If you are eating an average American diet, you should make the following changes to improve your health:

Dietary Recommendations:

Reduce fats (10)
Reduce simple sugars (10)
Increase complex carbohydrates (9)
Increase fiber (6)
Restrict salt intake (9)
Moderate alcohol consumption (10)
Maintain ideal body weight (10)

The numbers in parenthesis represent how many of the twelve official reports quoted, specifically made that recommendation. Now lets look at a few scientific studies which support these official recommendations.

The diets of over 850 people between the ages of 50 and 79 were carefully analyzed.[19] After twelve years of follow-up, the women who died of a stroke had eaten an average of 361 calories less per day than the women who did not die of a stroke. According to the simple energy balance theory, these survivors ate enough calories more than those who died of a stroke to have gained an average of over 35 pounds per year. In reality, the survivors actually weighed less than the stroke victims! (This paradoxical finding that those eating more weigh less shows up almost every time a study compares food intake and weight.) The stroke victims had eaten considerably less potassium than the survivors. The authors concluded that had those who died of stroke eaten just one extra serving per day of vegetables or fresh fruit, they would have improved their chances of survival by 40 percent.

A diet survey included over 8,000 Puerto Rican men aged 45 to 64 years.[20] Those urban men who died from coronary heart disease ate 153 calories per day less on the average than those who survived. Even more striking was the observation that those who died from other causes ate on the average 277 calories less per day than the survivors! The rural men showed the same trend; those who died of all causes ate an average of 246 calories less per day than those who survived. The survivors ate more carbohydrate than those who died, and especially more complex carbohydrate. The study suggests that eating more total calories and more complex carbohydrate protects people from dying of coronary heart disease as well as other diseases.

A study of Japanese men living in Hawaii showed that those who had heart attacks or died of coronary heart disease ate less carbohydrate, less vegetable protein, and 165 calories per day less than those without heart disease.[21] A study with brothers living in Boston or Ireland showed similar findings.[22] Those who died from coronary heart disease ate less fiber, starch, vegetable protein, and 147 calories less per day than those who did not die from coronary heart disease. A study from England placed men into three groups according to how many calories they ingested.[23] Those who ate the least had three times the incidence of coronary artery disease than those who ate the most. Other studies have also shown a protective effect against developing arteriosclerosis from eating more food from plant sources (vegetables, grains, and fruit).[24,25,26]

A number of studies have been done in which the diets of those with high blood pressure (hypertension) have been compared with those with normal blood pressure. High blood pressure has been associated with low intake of fiber,[27] potassium,[28,29,30] and calcium.[31,32] One study pointed out that those with hypertension were eating an average of 302 calories per day less than those with normal blood pressure.[33] The study concluded that "deficiencies rather than excesses are the principle nutritional patterns that characterize the hypertensive person in America."

A researcher was able to cause high blood pressure in rats by repeatedly restricting their food intake, alternated with allowing them to eat more.[34] This eating pattern is very similar to that of many people who frequently diet in an attempt to lose weight. That food restriction followed by eating may be a factor in causing high blood pressure is further demonstrated by the dramatic increase in hypertension after the seige of Leningrad during the Second World War,[35] and after other situations involving restricted food intake.[36,37]

Much attention has recently been focused on the relationship between diet and cancer, and especially the role of certain substances in food which may protect against developing cancer. A recent study from Greece compared the rate of breast cancer for five groups of women based on intake of vegetables. Those eating the least vegetables had ten times the incidence of breast cancer compared to those eating the most.[38] Other studies have also shown a protective effect from eating more fruits and vegetables.[39,40,41] Various studies have shown that certain vitamins and minerals protect against developing various cancers.[42,43,44] Other researchers have suggested that various other chemicals naturally found within foods protect against cancers.[45,46,47,48]

The studies above demonstrate the importance of complex carbohydrates in preventing high blood pressure, strokes, heart disease and cancer. The dietary recommendations to eat more complex carbohydrates and fiber are basically recommendations to increase vegetables, fruit, and grains. Reducing fats and refined sugars in the diet will automatically result in the intake of more vegetables, grains, and fruit if one eats enough food to satisfy hunger. Official dietary guidelines have not included guidelines regarding how much to eat. It seems clear from much of the information presented in this book that **eating plenty of nutritious food is a very important part of preventing serious diseases, preventing obesity, and for good emotional health and well being.**

There has been some very intriguing research which has shown that underfeeding certain experimental animals will cause them to live longer.[49,50,51] If they are restricted to the point that they are malnourished, or if underfeeding is alternated with free feeding, then they don't live as long. One study showed that if rats and mice were placed on a restricted food intake, and given free access to a saccharin solution, they had a much higher death rate than those animals who drank water.[52]

We are deeply concerned because some researchers are suggesting that people could prolong their lives by also eating less. There is no good long term evidence in humans that this is true, and much research to suggest that eating more is better. It is unjustified to draw these conclusions about humans from animals confined to small cages and fed animal chow. Most people in our society appear to be already deficient in various nutrients due to the highly refined nature of their diets. To cut nutrient intake even further could have disastrous consequences.

Now that you know what you can do to eat a healthier diet, you need to answer some important questions about the role of artificial sweeteners.

Do artificial sweeteners help you eat a healthier diet?

Most mammals seem to have a built-in mechanism that helps them know which foods have the needed nutrients and how much of each food to eat in order to stay healthy. Part of that built-in system involves a drive for sweet things. In the natural state, sweet-tasting foods generally have more energy than less sweet foods. This built-in desire for sweet things may be nature's way of assuring that each animal eats the kind of foods that will efficiently provide enough energy for its biological needs. Most animals also have a built-in drive for foods that are high in fat content, presumably because high-fat foods are also high in energy and will more efficiently fulfill energy needs. When given access to high-fat and highly sweetened foods, animals will generally prefer them—especially if the food is high in both fat and sugar.[53]

In an experimental situation in which animals are given their choice of a number of foods, each of which have essential nutrients, they choose enough of each food to remain healthy.[54,55] If they are given access to food for only one hour daily, they still eat enough to maintain their body weight, and choose enough of each essential food item to remain healthy. If a solution containing sugar is available during that hour, even though they have free access to water throughout the day, they will drink the sugar water to the exclusion of the essential nutrients, and will lose weight. After a few weeks, they will literally die of malnourishment.[56]

If animals are given a saccharin solution during the hour they can eat, they also choose the saccharin solution over the essential nutrients—but the results are not as drastic as with sugar. But here's the catch: a combination of artificial sweetener and sugar causes the greatest disruption of all.

Another interesting study evaluated the ability of experimental animals to regulate sodium. When rats had their adrenal glands removed, their kidneys lost extra sodium. If sodium was available to them, they would eat enough of the high-sodium food or drink enough of a high-sodium solution to maintain good sodium balance. Even when given access to a sugar solution, they were able to ingest enough sodium. If they were given access to a sugar solution a few weeks before surgery was performed, however, they developed a taste for it, and chose it over the substance they really needed—the salt solution. Under these circumstances, these animals actually died of sodium depletion![57]

Humans have also been shown to have the ability to eat enough of the right foods to keep them healthy,[58] but only if the food choices are from a ". . . 'primitive diet' based on natural foods, and not processed foods, such as sugar".[59] "A large body of evidence supports the existence of an innate preference for sweet-tasting substances in human infants."[60,61,62] This desire for sweet-tasting foods remains throughout life.[63] Mixing fats with sugar causes an even greater preference for these foods.[64]

From these and other studies, it is quite clear that highly sweetened foods and drinks can interfere with healthy food selection. A great deal of experimental work has been done in an attempt to identify the mechanisms by which mammals control their energy and nutrient intake. Some of the brain hormones involved in this process are discussed in Chapter Two. Although the entire process is not clear, some property associated with the sweet taste appears to be the main factor that causes this interference with healthy food selection. Since artificial sweeteners are much sweeter than sugar, under some circumstances, they can interfere even more than sugars!

Let's take a look at how artificial sweeteners affect each of the dietary recommendations:

Do artificial sweeteners help reduce fat intake?

From patients we have treated over the years, we believe that the use of artificial sweeteners is associated with eating more fats. Although there is not a lot of experimental work in this area, a few studies do confirm our observations. One controlled study showed that dieters using products sweetened with aspartame increased their fat consumption by 18 percent.[65] Another study showed a relationship between highly sweetened food and fat intake. There is a direct correlation between the amount of sugar eaten by people in various countries and the amount of fats those people eat. Countries where people eat few sugars are the countries where people eat few fats; those countries with high sugar use also had high fat intakes.[66]

There are a number of ways by which artificial sweeteners may increase fat intake:

1. Artificial sweeteners are well known to increase the appetite in general[67] and the desire for sweets in particular.[68] A great deal of the highly sweetened food available is also high-fat food. High fats are thus often eaten in response to a specific desire for sweets and in an attempt to satisfy the excessive hunger triggered by artificial sweetener ingestion.

2. Those who use highly sweetened food find that less sweet foods are not nearly as appealing. Not only do these foods not taste as good, they are not very satisfying, either. Because of this, they eat smaller amounts of vegetables, whole-grain products, and fruits when they are using highly sweetened products. The foods left to provide the needed energy are usually higher in fats and sugars, and these are usually eaten in larger quantities.

3. Those who make an effort to eat fewer calories seem to develop an increased desire for high-fat foods. Obese people especially seem to desire high-fat foods.[69,70] This increased desire for fats is particularly apparent in those who are restrained eaters.[71] The use of artificial sweeteners to help decrease food intake and to help increase the intervals between eating may be a major factor in an increased desire for fats.[72] Even though the total intake of fat may not be increased in a dieter using artificial sweeteners, the ratio of fats to other foods is usually increased.

4. Saccharin given to rats has been shown to cause a continual increase in levels of a certain type of opioid, which in turn has been shown to cause an increase in their preference for and intake of fat.[73]

Do artificial sweeteners help decrease sugar intake?

As was pointed out in detail in the last chapter, artificial sweeteners not only increase the desire for sweets, but usually contributes to an increased intake of sugars.

Do artificial sweeteners help increase complex carbohydrate and fiber intake?

There is currently a lot of discussion regarding the importance of complex carbohydrates and fiber to health. If you've been confused about these terms, a little clarification might help: Table sugar (sucrose) is called a simple sugar because it consists of only two sugar molecules linked together. They can quickly be broken down to single sugar molecules, which can then be easily absorbed into the bloodstream.

Complex carbohydrates consist of many sugar molecules linked together in chains, sometimes up to several thousand sugars long. Each sugar molecule must be split off from the ends of these long chains before they can be absorbed. These sugar compounds are considered to be complex carbohydrates for another reason. They are "complexed" within the food that contains them with a number of other substances. These substances, including fiber, slow down the digestion of these carbohydrates. In addition, a great number of vitamins, minerals, and other important food molecules are contained in these whole food products.

Fiber is the indigestible part of the plants that we eat for food. It serves many important roles, and will be discussed in more detail later. Fiber and complex carbohydrates are usually found in the same food sources. We would like to point out that sprinkling bran (which is high in fiber) on the top of ice cream or using bran in a breakfast cereal composed mostly of refined sugar and white flour is not the same thing as eating unprocessed food naturally high in fiber and complex carbohydrate.

Compared to refined sugars and artificially sweetened products, foods high in complex carbohydrates and fiber are not very sweet. Removal of the fiber seems to intensify the sweet taste of flour. White bread, therefore, is sweeter than whole wheat bread. That likely explains why people who eat highly sweetened food usually prefer white bread to whole wheat bread.

It is quite clear that the regular use of artificial sweeteners interferes with the taste, enjoyment, and satisfaction that could otherwise be obtained by eating foods high in complex carbohydrate and fiber.

Do artificial sweeteners help decrease salt intake?

Our patients often report a salt craving along with cravings for sweets. As long as highly sweetened food is regularly used, it seems that people enjoy their food highly seasoned and highly salted. When people first reduce both their sweetener and the amount of salt they use, food seems very bland and unappealing. After two weeks (and sometimes sooner), almost everyone who faithfully cuts down on sugars, artificial sweeteners, and salt intake will soon enjoy the new taste of these foods as much as ever. It has been our observation that people who try to cut back on salt while continuing to eat highly sweetened food maintain a strong desire for salt—and they continue to find low-salt foods unappealing.

Even though we can't find any research in this area, it is our clinical impression that the use of highly sweetened foods increases the desire for salt. There are several possible reasons why:

1. Highly sweetened food eaten on a regular basis may interfere with the taste mechanism, making it difficult to detect less potent flavors.

2. Most salt used in our society contains dextrose, a form of sugar. Those who desire highly sweetened foods may find highly salted food somewhat sweet, and be attracted to the sweetness.

3. Adequate levels of zinc are critical to the proper function of the sense of taste. In fact, one of the tests for zinc deficiency is a taste test. Solutions of several substances, including sugar and salt, are placed on the tongue in various concentrations. If you can detect these products only at the higher concentrations, then it is concluded that you are zinc deficient.

Zinc is found in small amounts in all unprocessed food—but none is present in sugar, and zinc content is reduced by processing various foods. People who use artificial sweeteners usually rely to a great extent on processed foods, and may eventually develop a subtle zinc deficiency, requiring higher levels of sweetness and salt in order for food to taste good.

Do artificial sweeteners influence alcohol intake?

We are not aware of any research that sheds any light on this question.

Do artificial sweeteners help you maintain an ideal body weight?

It is very clear that artificial sweeteners promote weight gain (as explained in Chapter One). Those who do manage to keep their weight under control while using artificial sweeteners usually do so only by constantly restricting the amount of food they eat. This brings up another important consideration related to good health:

Do artificial sweeteners interfere with obtaining adequate nutrients essential for good health?

It is now clear that one of the most important features of a healthy diet is getting enough calories to provide for your energy needs. In addition, you need enough essential nutrients to keep your brain and body functioning well, and to protect you from various diseases. We are firmly convinced that the use of artificial sweeteners, processed foods and the continued restriction of food intake for any reason is a major cause of health problems.

Artificial sweeteners may interfere with getting enough nutrients in several ways:

1. Artificial sweeteners encourage intake of refined foods and fats, both of which are nutrient deficient.

2. Artificial sweeteners promote weight gain. Since weight gain has traditionally been blamed on too many calories, most people who gain try to cut calories, and many go on extended reduced-calorie diets. This causes more weight gain, trains their bodies to live on less calories, and encourages them to eat more high-fat and high-sugar foods.

3. Artificial sweeteners seem to interfere with the ability to select foods that contain the nutrients your body needs. They also cause a decreased desire for unprocessed foods which

are high in essential nutrients, since these healthy foods are naturally less sweet-tasting.

Occasionally eating something sweetened with artificial sweetener will not turn you into a junk-food junkie. Some people use artificial sweeteners and still eat a healthy diet. But artificial sweeteners, especially if used frequently and in large amounts, interfere with the enjoyment and satisfaction obtained from healthy foods. Most people soon stop eating foods that don't taste good and that aren't satisfying. Naturally, they tend to gravitate towards the foods that appeal to them and that best satisfy their needs: the high-sugar and high-fat foods.

Of particular concern is the effect of artificial sweeteners on the diet of growing children. A wise pediatrician once taught that when you begin to introduce solids into a baby's diet, you should never begin with fruits. Why? The fruits are so sweet that, by comparison, the vegetables, cereals, and meats taste very bland. Once they acquire a taste for fruit, babies often won't eat the less sweet foods. That pediatrician's advice has proved true over the years. If the sweetness in fruit interferes with good food selection, then the much higher sweetness level associated with sugar and artificial sweeteners probably interferes even more with the eating of wholesome foods.

Many mothers who bring their children for medical treatment express a concern about not being able to get them to eat wholesome foods. Many children who are given access to highly sweetened food never do learn to like wholesome food. By the time they are teenagers, they often eat candy bars and soda pop for lunch—because those are the things they prefer over school lunch. Even adults who were allowed highly sweetened food as children often never learn to enjoy vegetables and whole-grain foods. In fact, a high percentage of the overweight people that come to us for weight control fit into this category, and probably wouldn't be nearly so heavy had they eaten more healthy foods instead of foods high in fats and sugars.

The bottom line?

Adequate intake of calories and protective nutrients is vital to optimal health. If you don't eat enough healthy foods, and eat too many unhealthy foods, you will begin to suffer from health problems. Highly sweetened foods sabotage your efforts to eat the way you should, and you suffer the consequences.

CHAPTER FOUR

ARE ARTIFICIAL SWEETENERS USEFUL FOR DIABETICS?

Maybe you've heard one of the most popular claims around: artificial sweeteners give diabetics the chance to have sweets without the problems caused by sugar.

Sound too good to be true?

To make a judgment, you first need to understand something about diabetes.

Diabetes (diabetes mellitus), simply stated, is a metabolism problem that results in too much sugar in the blood. There are two basic types of diabetes:

Type 1, or Insulin-Dependent Diabetes, occurs when too little insulin is produced. As described in Chapter One, insulin is essential if your cells are to take in and burn sugar for fuel. The cells in the pancreas that normally produce insulin may be damaged by infections, by antibodies, or by toxic substances. Since little or no insulin is being produced, insulin has to be given by injection. Many victims of Type 1 diabetes—the most serious type of diabetes—are young children.

Type 2, or Non-Insulin-Dependent Diabetes, occurs when insulin production is normal (or sometimes more than normal) but, for one reason or another, the body cells become resistant to the action of insulin. As a result, more insulin than usual is required to get enough sugar into the cells.

Your ability to produce insulin gradually decreases as you get older. Most people with Type 2 diabetes can produce enough insulin to keep their blood sugars reasonably normal for many years. Eventually, however, they just can't produce enough, the blood sugar levels climb above normal limits, and diabetes is diagnosed. Type 2 diabetes is a progressive disorder. It starts off with mild changes in the response to carbohydrates, and then becomes gradually worse. In addition to the decrease in insulin production over the years, there is also an increase in the resistance to the action of insulin.

In the early stages of Type 2 diabetes, the blood sugar increases rapidly after a meal, followed by an increased production of insulin, which often causes the blood sugar to fall rapidly to abnormally low levels a few hours later. This low blood sugar (or hypoglycemia) causes various unpleasant symptoms, and may be one of the first warnings of a diabetic tendency.

Type 2 diabetes is the most common type of diabetes, and can actually be improved or even reversed by appropriate treatment strategies. Unless specifically noted, further references to diabetes in this chapter will refer to Type 2 diabetes.

Since insulin resistance is the key defect in diabetes, it helps to understand the factors that contribute to or cause insulin resistance:

Factors Causing Insulin Resistance

1. **Antibodies** may form, and cause insulin resistance.[1] The antibodies may form against insulin itself,[2,3] or to the receptor sites on the cells to which insulin must attach in order to function.[4] These antibodies may interfere with the normal function of insulin, and thus interfere with sugar metabolism.

2. **Increased fat cell size** causes insulin resistance and is, of course, associated with obesity.

3. **Mineral deficiency** may contribute to insulin resistance. A number of minerals are important for insulin function (potassium, magnesium, calcium, copper, manganese, zinc), and if any of these are lacking, the action of insulin may be impaired.[5,6,7] The minerals come almost exclusively from the

diet, and if the diet is inadequate, mineral deficiencies may occur.

4. **Muscle inactivity** appears to be a major factor in insulin resistance. In fact, one of the most important benefits of exercise is an improvement in sugar metabolism and a decreased insulin resistance.

5. **Impairment in the cephalic phase of insulin release** may contribute to insulin resistance and diabetes.[8,9,10] As described in Chapter One, failure to produce an adequate initial surge of insulin in response to eating will interfere with the storage of sugar (glycogen) in the liver and will allow a rapid rise in blood sugar, resulting in an overproduction of insulin. Insulin overproduction is thought to be a major factor in the eventual development of insulin resistance.[11,12,13]

6. **Damage to certain areas of the brain—achieved by surgical cuts or toxic chemicals** in experimental animals—leads to a condition of insulin resistance. These defects (in the ventromedial hypothalamus (VMH) area of the brain) were described in Chapter One as a method by which obesity could be caused without increasing the number of calories.

There is a clear-cut relationship between obesity and diabetes: both are associated with insulin resistance. At least 80 percent of Type 2 diabetics are overweight.[14] Some experts have suggested that the insulin resistance and diabetes are caused by obesity and the associated large fat cells. Others believe that the obesity is caused by the diabetes and the associated insulin resistance.

We propose that the degree of insulin resistance is partly regulated by control centers in the brain. We know that control centers regulate weight and fat cell size. One of the mechanisms by which these brain control centers can control body fat storage is through controlling the degree of insulin resistance. If the control centers "choose" to store more fat, they merely need to increase the insulin resistance, more insulin will be produced to overcome this resistance, and the extra insulin will make you fatter. More of the sugar in your bloodstream is converted to fat, and the fat is then stored in the fat cells. The excess insulin also inhibits the breakdown of fat for fuel, even when you haven't eaten for a long time and are hungry and weak. Increased insulin and increased insulin resistance, then, will make you fat and keep you fat.

If your control centers "choose" to store great amounts of fat, a high degree of insulin resistance is needed to maintain the accompanying large fat cells. With increased insulin resistance, more insulin is needed in order to keep blood sugars at normal levels. If you can't produce enough insulin to compensate, then your blood sugar gets too high, and diabetes results.

Health Problems Associated With Diabetes

Although diabetes is mainly a defect of carbohydrate metabolism, fat metabolism is also adversely affected. These two defects both contribute to hardening of the arteries. Damage to small arteries and capillaries leads to eye damage and eventual blindness. In the kidneys, small vessel damage causes progressive kidney failure accompanied by high blood pressure and fluid retention. Damage to larger vessels greatly accelerates the rate of coronary artery disease, leading to heart attacks. Damage to cerebral vessels leads to a deterioration in brain function and to strokes.

That's just the beginning. Many other problems and symptoms result from the excessive levels of blood sugar and associated defects.

Goals For Dietary Management Of Diabetes

In treating both types of diabetes, the intent is to accomplish the following dietary objectives:[15,16]

1. Provide a diet that is nutritionally adequate—with enough fats, proteins, carbohydrates, vitamins, minerals, and other essential nutrients.

2. Achieve and maintain a desirable weight (and allow for growth in children).

3. Normalize, as far as possible, the metabolic abnormalities associated with the diabetes.

4. Prevent or delay the progression of the disease and its complications.

Traditional Treatment Of Diabetes

Before the availability of insulin, dietary changes were the only available treatment of diabetes. With Type 1 diabetes—since the blood sugar is too high and the sugar comes mostly from carbohydrate sources—the main treatment strategy was to reduce food intake, especially carbohydrate intake.

Long after insulin was available, treatment strategy for both types of diabetes still involved controlling food intake and cutting down carbohydrates. Food was limited to reduce the sugar levels and in an attempt to keep weight under control. Since the only major sources of food energy besides carbohydrate are fats and proteins, this meant an increase in the ratio of fats and protein in the diet.

Needless to say, the traditional diabetic diet caused a number of problems:

1. A high-fat diet contributes to hardening of the arteries. Since the diabetic state itself accelerates hardening of the arteries, the further insult of a high-fat diet is an absolute disaster!

2. A high-protein diet has been shown to cause hypertension in the small vessels of the kidney, leading to kidney damage and impaired function.[17] Since the diabetic state causes vascular damage to the kidneys, a high-protein diet may even be more dangerous in a diabetic.

3. A high-fat diabetic diet may worsen insulin resistance. Plant foods are the richest source of many essential minerals. Since a low-carbohydrate diet is usually low in plant materials, it is also usually low in certain minerals. There also seems to be something specific about a high-fat diet that directly increases insulin resistance.

4. A high-fat diet is associated with weight gain, even when no extra calories are eaten. Weight gain further increases the size of the fat cells, and makes the insulin resistance worse. Attempts to control weight through reducing calories causes further problems. Periods of sticking with the diet alternating with periods of going off the diet are frequently associated with more weight gain than would be a high-fat diet itself. A forced reduction in calories may cause these other problems:

a. Cutting back calories often causes loss of muscle and other protein tissues, and a gain in body fat. Even though the weight may not change, or may even go down, dieting can still lead to an increase in the amount of body fat.

b. Cutting calories leads to weakness, fatigue, and a decrease both in the ability and the desire to exercise. Almost no one who is on a calorie-restricted diet for very long will have the energy or ability to follow an effective exercise program, which is perhaps the most important available treatment for Type 2 diabetes.

c. Cutting calories may further deplete minerals vital to insulin function.

Improved Treatment For Diabetes

Recent research has improved our understanding and treatment techniques for diabetes. It now seems clear that diet has a significant impact on the development of diabetes. Some countries have a much lower incidence than others. In East Pakistan, the incidence of diabetes is only 2 percent, while in certain areas of the United States it is as high as 17 percent.[18] It has been thought that a high-fiber, low-fat diet protects against diabetes.[19,20]

The standard treatment for diabetes in a number of countries is a low-fat, high-complex-carbohydrate, high-fiber diet. In these countries, including Japan and India, there is a much lower incidence of vascular disease, heart attacks, and diabetic gangrene than among diabetic patients in the United States.[21] Diabetic patients in Japan also have a lower insulin requirement that those in the United States.[22]

For more than fifty years, some in the United States have advocated using a high-carbohydrate, low-fat diet for treating diabetes.[23,24,25] Dr. James Anderson has done twenty years of research using various diets for treating both Type 1 and Type 2 diabetics. He advocates a diet consisting of:

70 percent carbohydrate
11 percent fat
19 percent protein

He demonstrated that in all types of diabetes, this type of diet dramatically improved the level of insulin resistance. In many cases, those who had previously required insulin were under such good control that insulin was stopped completely. In other cases, insulin amounts could be decreased substantially.[26] These dietary recommendations closely parallel those of Nathan Pritikin, which have been also found to be very effective for treating diabetes.[27]

Unfortunately, it has only been recently that the work of these men has had some impact on the diet recommended by the American Diabetic Association, which in 1987 issued new dietary recommendations for diabetics. They suggested an increase in dietary fiber and a decrease in fat intake to less than 30 percent of the total calories with less than 300 milligrams cholesterol. Although these new guidelines are a step in the right direction, some researchers believe they don't go far enough.

A high-complex-carbohydrate, high-fiber, low-fat diet has some clear advantages over older American Diabetes Association diets. To simplify things—and as a tribute to the work that Anderson did—we'll call it the Anderson diet. Its advantages include the following:

1. The Anderson diet gives better control of diabetes. Insulin or other diabetic medication can often be reduced or eliminated.

2. The Anderson diet corrects some of the metabolic problems associated with diabetes. Sugars are absorbed more slowly and evenly. There is a more gradual rise in blood sugar after eating, and the blood sugar levels do not go as high. Insulin resistance seems to be lowered.

3. The Anderson diet seems to help reduce the development of arteriosclerosis, thus reducing or delaying the onset of many diabetic complications. While these dietary principles are healthy for everyone, they are particularly important for diabetics, who have a greatly accelerated rate of arteriosclerosis.

4. If used in conjunction with appropriate exercise, the Anderson diet is effective for weight control. This combination will not only reduce the amount of fat and the size of the fat cells,

but may at the same time cause an increase in muscle mass in those who have lost muscle from previous unhealthy diets.

Most Type 2 diabetics should be able to eat until completely satisfied on the Anderson diet. This has some major advantages, including the possible lowering of the weight setpoint in the brain and development of a more normal cephalic insulin response to eating. In addition, eating to satiety eliminates a lot of very unpleasant symptoms associated with inadequate food intake. Type 1 diabetics may have to follow some caloric guidelines, since their hunger control may be abnormal due to the unnatural way in which their insulin needs to be taken.

5. The types of foods recommended in the Anderson diet, if eaten in adequate amounts, should greatly improve energy, endurance, and strength. It should not only improve the ability to exercise, but should help people to feel good during and after exercise. The extreme importance of exercise in weight control and in improving the metabolic problems associated with diabetes has been mentioned.

   Many doctors have been unsuccessful in getting diabetic patients to exercise, so they have given up—and they no longer even advise patients to exercise. Why have diabetic patients been so stubborn about exercise? Because their doctors have usually prescribed a reduced-calorie diet at the same time. Simply put, they didn't have the energy to exercise. Proper instructions on the importance of exercise and the steps to take in getting started are important in getting patients to exercise. The real key, however, is getting them to eat enough calories so they'll have the necessary energy to exercise effectively.

6. Adequate amounts of food on the Anderson diet should provide all the minerals and other nutrients essential for improved insulin response, better sugar metabolism, and good nutrition in general.

The Use Of Sugar (Sucrose) By Diabetics

Although there have been a number of healthy trends in the recent treatment of diabetes, one trend concerns us greatly. A few experimental studies compared refined sugar to other types of carbohydrates. Under certain circumstances, it was found that other carbohydrates caused a rise in blood sugar and an insulin response just as high as that caused by sugar. Researchers have evaluated many foods for their blood sugar and insulin response, and have formulated a comparative figure called the glycemic index.

On the strength of these studies, some have suggested that sugars are probably okay for diabetics, if used in moderation. Although sugars appear to be no worse than some other carbohydrates if you look at a glycemic index, there are many other problems associated with sugar use by diabetics.

As has been described, adding sugar to the diet frequently causes a substantial weight gain, even if no extra calories are added. In most cases when sugar is added, there is an associated increase in hunger and in the desire for sweets—and people eat more. This makes the weight problem even worse.

Any refined sugar that is added to the diet displaces some other food source. Complex carbohydrates contain fiber, minerals, and other substances that can actually improve diabetic control in the long term. By using sugar, then, people not only make their diabetes worse, but they deprive themselves of the foods that could improve them.

Sugar has been shown to increase both triglycerides and cholesterol in diabetics.[28] This speeds up the rate at which arteriosclerosis develops. This problem and a number of other problems specifically associated with sugar use will be discussed in detail in Chapter Six.

The problems associated with sugar result from its refined nature, its sweet taste, and its lack of essential nutrients. Refined white flour and other highly refined products cause similar problems, although not to the same extent.

Every expert—even the scientists hired by the sugar producers—agrees that excessive sugars are a problem for diabetics. The disagreement is in how much sugar is acceptable. Since sugar is a problem, artificial sweeteners have been promoted as an excellent substitute. The public has bought the story: 72 percent of mothers of diabetic children believed that artificially sweetened foods helped their children stick to their diets.[29] Some claim that artificial sweeteners give all the enjoyment and benefits of the sweetness, but without the problems caused by sugar.

Let's take a look at that claim.

Are Artificial Sweeteners Useful For Diabetics?

Since specific management goals have been outlined, let's look at each to see how well artificial sweeteners help diabetics:

Do artificial sweeteners help diabetics eat a nutritionally adequate diet?

One study pointed out that diabetics using artificial sweeteners were not able to follow their diets any better than diabetics who did not use artificial sweeteners.[30] We have already pointed out that artificial sweeteners actually increase the desire to eat sugar—and that artificial sweeteners cause people to eat more sugar. We have also pointed out that artificial sweeteners interfere with the enjoyment and satisfaction obtainable from healthy foods, and thus interfere with the selection of a good, nutritious diet.

Do artificial sweeteners help diabetics achieve and maintain an ideal weight?

Research in this area has established that artificial sweeteners do not help with weight control—in fact, in most cases they seem to make weight problems worse. Diabetics, especially Type 2 diabetics, may be even more prone to gaining weight from the use of artificial sweeteners than are non-diabetics due to the effect on insulin. (For details, see Chapter 1.)

Do artificial sweeteners help normalize the metabolic problems associated with diabetes?

The main problems in diabetes are related to insulin metabolism. As pointed out, artificial sweeteners may interfere with the cephalic phase of insulin release, which makes blood sugar control worse. They may also encourage the type of diet that increases insulin resistance, making the metabolic problems associated with diabetes even worse.

Some studies done with diabetics using aspartame concluded that aspartame had no adverse effects on diabetes.[31,32] These studies were usually done using capsules containing aspartame; the capsules were swallowed, giving the diabetics no chance to taste the aspartame. Remember—most of the metabolic effect from artificial sweeteners comes from the sweet taste, so these studies can't correctly evaluate metabolic effects.

The insulin required by some diabetics may cause further problems with the amino acid metabolism and neurohormonal balance. Insulin injections in those using aspartame may further increase the blood levels of phenylalanine and decrease blood levels of tyrosine.[33] The symptoms caused by neurohormonal imbalance resulting from the use of aspartame may be worse in diabetics, or may be caused at lower doses than in non-diabetics.

Do artificial sweeteners help delay the progression of diabetes and its complications?

Since artificial sweeteners do not help—and, in fact, probably interfere with—the metabolic problems of diabetes, they probably speed up the progression of this disease. The complications of diabetes are influenced greatly by diet. Since artificial sweeteners interfere with choosing a good diet, they probably intensify diabetic complications as well.

From a health viewpoint, artificial sweeteners have nothing useful to offer for a diabetic, and seem to cause only problems. Some advocates of artificial sweeteners have argued that life is tough enough for the diabetic without being deprived of the pleasures from using sweeteners. They say that the sweetness obtained from artificial sweeteners will not hurt, and will help diabetics satisfy their desires for sweets, making it easier to tolerate the restricted diet. That argument makes as much sense as encouraging a Catholic priest to use pornography to help suppress his sex drive!

CHAPTER FIVE

ARE ARTIFICIAL SWEETENERS SAFE AND FREE FROM SIDE EFFECTS?

Are artificial sweeteners safe?

The answer you get depends on whom you ask!

The Food and Drug Administration (FDA) is charged with the responsibility of assuring that any new food additive or drug is safe. If artificial sweeteners were classified as drugs, they would not only have to be proven safe, but would also have to be proven to be effective before they would be acceptable. We wouldn't have artificial sweeteners, then—because they are clearly not effective for the purpose they are used.

Since medical claims or benefits can't be made for food additives, they only have to be proven safe. Since 1958, exhaustive safety testing has been required before any product passes FDA approval. Since saccharin was discovered in 1879 and cyclamates in 1937, the approval process was much more relaxed for these products.

The FDA is not concerned with side effects.

Virtually all drugs cause side effects in at least some people. Most drugs have some dangers, and many drugs can cause death in some situations. If the potential benefit of these dangerous drugs exceeds the risks, then they are approved in spite of the dangers.

Many food additives also have side effects. Monosodium glutamate (MSG), for example, has been linked with a frequent incidence of headaches and other problems.[1,2,3] Sulfiting agents have been associated with allergic reactions.[4,5,6,7] Because of at least fourteen deaths linked to sulfite-treated lettuce in salad bars, sulfites have recently been banned from use on fresh vegetables.[8] They are still used, however, in other ways.

Since virtually all drugs, foods, and food additives are capable of causing allergic reactions in a sensitive person, the FDA does not concern itself with allergic reactions. It does not seem to be concerned about the issue of side effects, either, even if they tend to be rather severe. They seem concerned only about products that cause actual structural changes or tissue damage.

Safety of Artificial Sweeteners

Although cyclamates were originally approved as being safe, a number of subsequent studies showed that experimental rats developed bladder cancer when given prolonged, heavy doses. As a result, they were removed from the market in the United States in 1969, although they continue to be available in many other countries.

Saccharin has also been implicated in many studies as a cause of bladder cancer or pre-cancerous lesions in experimental rats.[9,10,11,12,13,14,15] We don't yet know whether it causes or contributes to cancer in humans. In 1977, the FDA announced plans to ban saccharin. Because of public pressure, Congress put a moratorium on the ban, which has been renewed regularly since that time. The FDA agreed to a warning label on products that contain saccharin, alerting people to the fact that saccharin may be a cancer-causing agent.

Because of the experience with cyclamates and saccharin, and because of current strict controls, aspartame was studied extensively before it was approved. By July 1981, the FDA Commissioner stated that "aspartame had become the most heavily tested food product in history . . .," and concluded that ". . . all reasonable questions about the product's safety had been satisfactorily answered."[16] Aspartame was approved at that time for use as a tabletop sweetener, in mixtures for instant beverages, and as an ingredient for other foods that did not require cooking. In July, 1983, it was approved for use in soft drinks.

The use of aspartame rapidly skyrocketed, and some people were consuming it in far larger quantities than had been anticipated by the producers and by the FDA when the original research was done. Alarming reports of serious side effects began surfacing all over the country, and thousands of formal complaints were made to the FDA and to the Centers for Disease Control in Atlanta.

A number of scientists, consumer groups, and individuals have become very concerned about the health problems associated with aspartame use. A support group called "Aspartame Victims and Their Friends" has been formed to help those who have suffered damage from the use of aspartame. Various groups have encouraged the FDA to investigate the aspartame issue. The Community Nutrition Institute (CNI) recently petitioned the FDA to ban the use of aspartame as an "imminent health hazard" and to hold a public hearing. This group presented many cases of seizures and visual problems which they believed to be clearly associated with aspartame use. On November 21, 1986, this petition was denied.[17]

FDA Associate Commissioner for Regulatory Affairs, John M. Taylor, wrote to the CNI lawyer a letter that is very enlightening regarding their handling of these complaints. Taylor's letter states that "there was no consistent or unique pattern of symptoms with respect to aspartame intake that would indicate a causal link between the symptoms and aspartame consumption." In other words, to prove an association, symptoms had to be "consistent" and follow a "unique pattern." Did he mean that *everyone* who used aspartame had to develop symptoms before aspartame could be blamed for the symptoms? By "unique pattern" did he mean that the symptoms associated with aspartame had to be caused only by aspartame, and not known to be caused by anything else? He also stated that "many of the symptoms reported to be attributed to aspartame consumption are of a common nature." Is he implying that any symptom that is common couldn't be associated with aspartame use?

Regarding the matter of seizures being caused by aspartame, Taylor indicated that seizures were very common, and that approximately 1 percent of the population suffered from them. He also stated that "seizure susceptibility can be increased by a number of factors," and that "there may be a chance occurrence of seizure activity following ingestion of aspartame by seizure-prone people." He also dismissed the issue by stating that "such a happenstance, however, is not indicative of a causal relationship between consumption of aspartame and the onset of seizure activity."

Taylor indicates that 152 cases of visual problems have been investigated by the FDA. In all these cases, the people making the complaints were convinced that these problems were caused by aspartame. How did Taylor account for these complaints? Thirty-one of these cases were dismissed because "they were not accompanied by complete medical records. One-third of the reports reveal no association with aspartame consumption other than a temporal one." In other words, the problems came on during the use of aspartame, and disappeared when they stopped using it. Since there was no other proof that the aspartame caused the problems, these cases were dismissed!

"Approximately 30% of the reports described the symptom 'blurred vision' in connection with headaches. None of these cases appears to represent true visual problems," Taylor wrote. In other words, since headaches are often associated with blurred vision, these cases were dismissed. Since headaches are one of the most frequent symptoms associated with the use of aspartame, and many of these headaches were likely also caused by aspartame, it seems unjustifiable to dismiss these cases. "Eighteen reports of visual disturbances were associated with concurrent medications. Moreover, five reports were associated with the use of medications known to cause side effects such as blurred vision."

Five cases were dismissed because people were using a medication known to cause blurred vision. Even if these people had been taking these medications for years, and the blurring only happened when they took aspartame, the simple fact that they used the medication excluded them from consideration. Even more incredible was the dismissal of eighteen cases because the people involved were taking medication—*even though the medications were not known to cause blurred vision!* The remaining three reports were dismissed on the basis that the people involved were diabetics—and, as everyone knows, diabetics can have visual disturbances.

Here is the dilemma. Many cases of severe problems thought to be caused by aspartame have been presented to the FDA. The FDA easily rejects each one of them—either claiming that there is not enough medical evidence to rule out other causes, or that there is medical evidence to suggest that other factors could explain the side effects. According to the FDA, any medication, any other medical condition, or any other associated symptom is reason enough to reject the claims.

Taylor concluded that the data submitted "do not demonstrate a link between aspartame consumption and either the onset of seizures or the occurrence of eye damage and do not raise a reasonable question about the safety of aspartame consumption."

Have all the consumer complaints and the submission of claims by various groups opened the door for the FDA to consider future research? According to Taylor, ". . . Only well-controlled clinical trials which focus on specific endpoints would provide evidence for the existence of an effect in small populations of individuals with a reported response. . . . Because no evidence has suggested a consistent pattern of adverse effects (including seizures), there is no basis upon which to design clinical studies." The FDA not only won't accept the idea that aspartame can cause problems, but simply will not even consider the idea of doing further research.

Even though the FDA won't admit that aspartame causes problems, there is a great deal of evidence that it does create a wide variety of symptoms. Most of the proof comes from people who describe their own experiences with aspartame. Some scientists call this "**anecdotal evidence**," and don't believe that it is of much value in establishing scientific facts. But if many people report problems while using a product, and report that those problems disappear after stopping use, then that has to be meaningful information. Some of these reports may be just coincidental, but when enough people report the same symptoms, there surely must be some validity to the relationship.

In our medical practice, we have seen dozens of cases in which aspartame appeared to directly cause symptoms. We have seen literally hundreds of other cases in which aspartame was suspected, but could not be proved. In these cases, patients gave up artificial sweeteners—but, at the same time, they made improvements in their diet and also gave up sugar, chocolate, and caffeine. Clearing of symptoms is often very dramatic with this type of treatment, but it is impossible to determine which of the removed substances caused the original problems.

Not a single one of the cases we have seen were reported to the FDA or Centers for Disease Control (CDC). For every case reported, there are probably hundreds or thousands of cases that are not reported. For every case in which symptoms are linked to aspartame, there may be dozens more in which problems are caused by aspartame, but which go undiagnosed.

Features of problems caused by aspartame.

What kinds of problems are encountered in medical practice when patients use aspartame?

To answer that question, we'll use the cases of two patients from our practice who described their stories within the last month. They illustrate some important points.

Anne is a white female in her mid-forties who had a number of health problems. She finished her Ph.D degree in psychology, remarried, moved, and assumed a new teaching position, all within a short period of time. She began having severe headaches, which became progressively more frequent and severe. She also had periods of time when she had trouble seeing, and she couldn't even read at these times. An eye specialist prescribed new glasses, which didn't seem to help, and reassured her that her eyes were not causing the headaches. The headaches were improved a little with hormone (estrogen) treatment, and became worse during periods of stress.

Anne started feeling extremely anxious and fearful about teaching classes and presenting papers to professional groups. She began to doubt her own competency, and felt even more fear that she might not be able to do an adequate job with her new position. She began needing medication to sleep. She experienced a very distressing feeling of things crawling under her skin.

She developed severe pains in her left shoulder and neck, which were thought to be tendonitis or nerve root irritation.

All these complaints were thought to be related to stress and menopause. She started using narcotic painkillers in pills and by injection. She felt that the painkillers further interfered with her memory, which was becoming an increasingly severe problem. Hormone shots, tranquilizers, anti-inflammatory medications, and sleeping pills gave some relief.

When Anne was seen, it was explained that caffeine, sugar, and chocolate sometimes caused these problems, and suggested she stop using these things entirely. She did so, but no change occurred. She then found out that a friend got headaches from using aspartame, and wondered if that could be her problem. When she was questioned, she admitted that she had begun to drink huge amounts of diet cola when she was finishing her thesis, and she had continued to do so. She at times drank as many as eight sixteen-ounce bottles of diet pop *every day*. She felt that these helped her eat less, as she struggled unsuccessfully to control her weight. She knew she had a strong craving for aspartame products, but saw it as a harmless craving, since very few calories were involved.

She stopped using products sweetened with aspartame, and the results were remarkable. Within a few days of stopping aspartame, her headaches became less severe. Within a few weeks, they stopped almost entirely. Her anxiety lessened, and the crawling sensation under her skin disappeared. She started sleeping better, and her fears and feelings of inadequacy disappeared. In short, she felt just like she had before she began using aspartame. All the reasons and explanations previously given to explain her problems were simply not correct: her problems were almost without question caused by aspartame.

In the second case, Michelle began using large quantities of a powdered lemonade drink sweetened with aspartame because she received dozens of introductory coupons for it. She enjoyed the lemonade, and began buying other products for her family that were sweetened with aspartame. She believed that drinking lots of the lemonade throughout the day helped her cut calories, and she continued to diet, as she had throughout much of her life. She had suffered occasional headaches before, but began getting more severe headaches more often. Soon, she had almost constant headaches every day, and sometimes they became almost intolerable.

The headaches were only the beginning of Michelle's problems. Her hands started going numb. She began to notice problems with being unable to think clearly, and her memory became poor. In spite of writing things down, she would forget important appointments. She mentioned that she had reminded herself all morning of an important meeting she was to attend that afternoon, and the next thing she realized, it was late afternoon—and she had missed the meeting. She began to question her sanity, and became very anxious and fearful of "losing her mind."

A friend suggested that her problems might be caused by aspartame, and she stopped using it immediately. After one day, her headaches were completely gone. It took several months, however, before her memory and other problems returned to normal. A few months after she had stopped using aspartame, she had one diet pop containing it—and she developed a severe headache within minutes.

Michelle's case points out another interesting thing. In spite of the fact that she ate very little for months, she lost very little weight. Within five months of stopping the diet drink and stopping her starvation diet regime, she gained fifty pounds—a typical development following a very severe calorie-restricted diet.

These two case histories present a number of typical features related to aspartame usage and symptoms:

1. **Delayed onset of symptoms.** Symptoms may not develop until aspartame has been used for a number of weeks or even months.

2. **Gradual increase in symptom severity.** Symptoms may start at a very low level, and gradually worsen over the course of many months.

3. **Gradual resolution of problems after aspartame is stopped.** Many symptoms may not clear up completely for a number of weeks.

4. **Tendency to "abuse" aspartame.** Many people use aspartame products like addictive substances. They use it daily (often many times daily), crave it, gain great satisfaction from

using it, gradually increase their consumption, use it in place of food, and often take it in very large quantities.

5. **Symptoms tend to be common problems.** Since the type of problems reported from aspartame use are problems commonly encountered by a medical doctor, the cause is often overlooked.

Although a good share of problems from aspartame seem to be associated with high doses over long periods of time, some reactions may be immediate and following a small dose, as seen in Michelle's headaches.

Let's now look more closely at some of the symptoms most commonly reported with aspartame usage.

Symptoms associated with aspartame

Dr. H. J. Roberts has documented a group of 496 people who reported problems from using aspartame. The following list (Table 1) gives the type of symptoms reported, and the percentage of this group who reported each symptom.[18]

HEADACHES are the most common complaint associated with aspartame reported to the FDA.[19] Migraines caused by aspartame have been referred to in the medical literature.[20] Migraine and other headaches are well known to be caused by foods and food additives.[21,22,23,24,25,26,27]

SEIZURES have been repeatedly reported in association with aspartame usage.[28,29] Some studies in animals have shown that aspartame can contribute to seizures.[30] Between those cases reported to the FDA and those reported to Richard Wurtman and other doctors, several hundred cases of seizures related to aspartame use are under investigation.[31]

Seizures are serious problems. Imagine having a seizure in the supermarket—or, worse, having a seizure and losing control of your bowels and bladder in public. Worse yet, imagine having a seizure and losing control of your car on the freeway. Imagine not being able to drive for at least a year, and the inconvenience that this might cause to you and your family.

While not pleasant for anyone, seizures can be an especially devastating problem for some people. Dr. Wurtman describes an Air Force pilot stationed at an air base in the desert. He began

TABLE I

MAJOR SIDE EFFECTS IN 496 ASPARTAME REACTORS

Eye
Decreased vision and/or other eye problems (blurring, "bright flashes," tunnel vision)	118	(24%)
Pain (one or both eyes)	44	(9%)
Decreased tears, trouble with contact lens, or both	12	(2%)
Blindness (one or both eyes)	12	(2%)

Ear
Tinnitus ("ringing" or "buzzing")	63	(13%)
Severe intolerance for noise	38	(8%)
Marked impairment of hearing	23	(5%)

Neurologic
Headaches	225	(45%)
Dizziness, unsteadiness, or both	195	(39%)
Confusion, memory loss, or both	140	(28%)
Convulsions (grand mal epileptic attacks)	73	(28%)
Petit mal attacks and "absences"	18	(4%)
Severe drowsiness and sleepiness	81	(16%)
Paresthesias ("pins and needles," "tingling") or numbness of the limbs	68	(14%)
Severe slurring of speech	55	(11%)
Severe "hyperactivity" and "restless legs"	39	(8%)
Atypical facial pain	33	(7%)
Severe tremors	41	(8%)

Psychologic—Psychiatric
Severe depression	125	(25%)
"Extreme irritability"	109	(22%)
"Severe anxiety attacks"	86	(17%)
"Marked personality changes"	76	(15%)
Recent "severe insomnia"	64	(13%)
"Severe aggravation of phobias"	32	(7%)

Chest
Palpitations, tachycardia (rapid heart action), or both	72	(15%)
"Shortness of breath"	42	(9%)
Atypical chest pain	34	(7%)
Recent hypertension (high blood pressure)	23	(6%)

Gastrointestinal
  Nausea   68 (14%)
  Diarrhea   63 (13%)
    Associated gross blood in the stools   12 (2%)
  Abdominal pain   62 (13%)
  Pain on swallowing   22 (4%)

Skin and Allergies
  Severe itching without a rash   38 (8%)
  Severe lip and mouth reactions   23 (5%)
  Urticaria (hives)   21 (4%)
  Other eruptions   42 (9%)
  Aggravation of respiratory allergies   10 (2%)

Endocrine and Metabolic
  Menstrual changes   40 (8%)
    Severe reduction or cessation of periods   18 (4%)
  Marked thinning or loss of the hair   31 (6%)
  Marked weight loss   22 (4%)
  Paradoxic weight gain   29 (6%)
  Aggravated hypoglycemia (low blood sugar)   20 (4%)
  Loss of control of diabetes   15 (3%)

Other
  Frequency of voiding (day and night), burning on 60 (12%)
    urination (dysuria), or both
  Excessive thirst   52 (10%)
  "Bloat"   50 (10%)
  Severe joint pains   48 (10%)
  Fluid retention and leg swelling   18 (4%)
  Increased susceptibility to infection   6 (1%)

Table I. Clinical symptoms attributed to aspartame by Dr. H.J. Roberts (A clinician's adventures in medicine: Is aspartame safe? On Call (The official publication of the Palm Beach County Medical Society) January 1987; pp. 16-20.

drinking huge quantities of diet pop sweetened with aspartame to replace the fluids he lost from running in the desert heat. After a few months of this, he suddenly had a grand mal seizure. It destroyed his career—any pilot who has a seizure for any reason is automatically grounded, and will never be able to pilot a plane again.

A few days after hearing this incident, a pilot came to our office for some health problems. He had been an Air Force pilot, and had then worked for United Airlines for several years. He mentioned being on medical disability since having a seizure two years earlier. He had been given an extensive medical workup, and no cause for the problem had been identified. Upon close questioning, he mentioned that he had started to drink diet pop sweetened with aspartame a number of months before the seizure. He also mentioned that his wife had been on a sugar-avoidance campaign, and had been buying a lot of desserts, diet gelatin, hot chocolate mixes, and other products containing aspartame. When we added it all up, he had been using quite a bit of aspartame.

Was aspartame responsible for ruining the careers of these two pilots? We'll probably never know for sure. It is certainly likely, however, that aspartame was a major contributor—and that without it, they may have never experienced a seizure.

Dr. Wurtman suggested that headaches and seizures may go together. Headaches may gradually worsen, and the victim may eventually have a seizure.[32]

Possible causes of side effects.

You need to understand a little about aspartame metabolism in order to understand the cause of its side effects. Aspartame is made from two amino acids (phenylalanine and aspartic acid), which are linked together with a molecule of methanol (methyl alcohol). Quickly after you eat it, aspartame breaks down completely into these three ingredients, and they are processed separately by your body. The two amino acids may be used as building blocks for the formation of various proteins. Each product may be broken down through a number of different metabolic pathways. (See figure 7.) Side effects may be related to these products or their metabolic breakdown products.

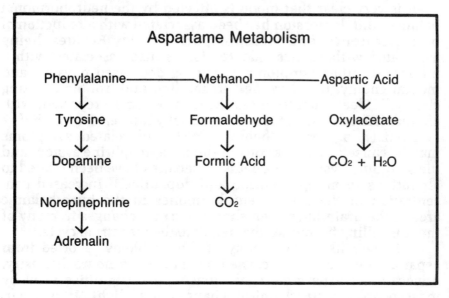

Figure 7. Structure of aspartame molecule, and the breakdown products produced by its metabolism.

Some of the possible mechanisms include the following:

1. **Neurotransmitter Balance.** It is clear that large doses of aspartame change the ratio of amino acids in the bloodstream[33,34,35] and the balance of various neurochemicals in the brain.[36,37,38] Aspartame seems to lower blood levels of the amino acid tryptophan and lower brain levels of serotonin. Aspartame is well known to cause increased blood levels of phenylalanine. Although this situation causes higher levels of tyrosine in the blood and brain of the rat, it actually causes lower levels of tyrosine, dopamine, norepinephrine, and adrenalin in the human brain.[39] (Humans convert a much smaller proportion of phenylalanine into tyrosine than does the rat.)[40] Since phenylalanine competes with tyrosine to get into the brain, increasing the level of blood phenylalanine actually decreases the level of brain tyrosine, and the neurohormones derived from it.

It is very clear that mood is affected by the neurohormonal balance, and depression has been associated with a reduction of both serotonin and norepinephrine.[41,42,43] Besides being associated with sadness, depression is also associated with a number of other symptoms, including decreased physical and mental energy, mood swings, irritability, and trouble thinking clearly.[44] Sleep disturbances have also been reported with altered neurochemical balance, especially low serotonin.[45,46,47]

Anxiety, aggressive behavior[48,49,50] and related symptoms may be caused by alterations in the norepinephrine system and the serotonin system.[51] Convulsive seizures have been related to alterations in norepinephrine and dopamine.[52] Increased concentration of these two neurohormones in the hypothalamic area of the brain have been shown to cause changes in many of the controlling hormones that regulate the menstrual cycle.[53,54]

It is very likely that many of the problems reported from aspartame use could be caused by changes in neurochemistry. Advocates of aspartame claim that in the quantities used by most people, neurochemical changes are slight. It is clear, however, that even small changes in neurochemistry—such as the changes caused by the simple act of eating—can result in significant changes in mood, hunger, and behavior.[55,56,57,58,59]

The use of amino acid supplements to improve certain neurohormonal levels and improve symptoms is good evidence that neurohormonal imbalance is a cause of depression, sleep disturbance, and some of the other problems mentioned above. L-tryptophan supplements have been used successfully to improve sleep disturbances[60,61,62] and depression[63,64,65,66] by improving brain levels of serotonin. Tyrosine has been shown to increase brain levels of dopamine and norepinephrine[67,68,69] which in turn has had some effect in improving depression.[70,71,72] Of further interest is the observation that using tryptophan and tyrosine together has been more effective for treating depression than either used by itself.[73]

An interesting effect occurs when tyrosine supplements are given to rats: it lowers their blood pressure.[74] Since aspartame has the opposite effect in humans (reduces tyrosine levels in the brain) it may contribute to high blood pressure in them.[75]

The gradual onset of symptoms with aspartame use, and the gradual clearing of symptoms when it is stopped, is very typical of medical conditions caused by neurohormonal imbalance. When antidepressant medications are used to alter neurochemical levels to control depression, improvement may not occur for ten days to two weeks—and it may take a number of additional weeks before maximum improvement occurs.

2. **Allergy.** Some of the symptoms reported from aspartame use may be from an allergic or adverse reaction to the aspartame or one of its metabolic breakdown products. **Allergy** is defined in Webster's Dictionary as "a condition of unusual sensitivity to a substance or substances which, in like amounts, do not affect others."

At one point, allergists identified one antibody system (IgE) that caused the typical skin reaction and produced some of the typical responses such as hives, eczema, hay fever, and asthma. Some allergists narrowed the definition to include only those responses caused by IgE antibodies. Since then, a number of other antibody systems and other substances that contribute to adverse reactions have been identified. For simplicity, we would like to refer to all of these unusual reactions—regardless of the cause—as allergic, in keeping with the original definition.

Allergic reactions may cause a wide range of symptoms in addition to the classical reactions of hives, eczema, hay fever, asthma, and anaphylactic shock. Migraines and other headaches, dizziness, weakness, anxiety, panic attacks, depression, blurred vision, trouble thinking clearly, sleepiness, body aches, arthritic pains, digestive disturbances, elevated blood pressure, and changes in blood sugar are all well-known adverse reactions.

Some people are very sensitive, and can react adversely to even tiny amounts of a substance and to a wide range of things. Others react only when a substance gets above a certain level. Although IgE reactions have been thought to occur mainly to proteins, it now appears that many allergies can occur to various chemical substances.

The metabolic by-products of aspartame digestion include the following substances. Although some of them are "natural," and may be found in common foods, large amounts may exceed the tolerable level and lead to reactions:

Aspartame
Beta-aspartame
Aspartic acid
Phenylalanine
Methanol
Formaldehyde
Formic Acid
Aspartylphenylalanine
Aspartylphenylalanine diketopiperazines
Aspartylphenylalanine amide
Tyrosine
L-Dopa
Dopamine
Norepinephrine
Epinephrine
Phenylethylamine
Beta-phenylethylamine
Phenylpyruvate
Phenyllactic acid
Phenylacetic acid

These products are all broken down further, accounting for other metabolic by-products in larger amounts than usual. In addition, increased amounts of various enzyme systems occur in response to large amounts of aspartame.

Whenever a substance is refined or synthesized, impurities may result from the process. Petrochemical residues are found, for example, in sucrose (table sugar) and saccharin. Although the residues are not found in large amounts and are not thought to be toxic at those levels, people sensitive to petrochemical products may react badly to these impurities.

Some of the by-products of aspartame metabolism are now well known to cause allergic reactions. Formaldehyde has been widely studied, and a number of reports of allergic reactions are reported in the medical literature (see references 104 to 107). Those who react to new carpets, new clothing, and to chemical odors from new buildings are probably sensitive to formaldehyde.

Aspartic acid is very closely related to glutamic acid (glutamine), which is well known to cause reactions. Monosodium glutamate (MSG), the cause of "Chinese Restaurant Syndrome," causes adverse reactions in up to 25 percent of people who eat it. MSG was approved for many years before researchers discovered its tendency to cause headaches and other adverse reactions.[76] It seems probable that aspartame will also prove to be a factor in allergic reactions. Until aspartame came along, there was no common way in which aspartic acid was used in a way that its levels would become higher than other amino acids.

3. **Blood Sugar Effects of Artificial Sweeteners.** The tendency of artificial sweeteners to affect blood sugar was discussed in detail in earlier chapters. By interfering with the cephalic phase of insulin, there is a tendency for the blood sugar to fluctuate widely among people who use artificial sweeteners on a regular basis. Low blood sugar or rapidly falling blood sugar may be associated with fatigue, weakness, trouble thinking clearly, and trouble staying awake. The stress hormones produced in response to falling blood sugar may cause symptoms like anxiety, irritability, palpitations, and trouble sleeping. Drastic changes in blood sugar have been associated clinically with mood swings, variable energy level, and headaches.

4. **Starvation Effect.** Extensive studies have been done on people during periods of limited food intake. During World Wars I and II, various groups of prisoners-of-war, civilians with food rationing, and experimental subjects were studied.[77] All groups reported very similar symptoms, even when the food was not that limited.[78] Some of these subjects were eating up to 2,200 calories daily. In fact, many of these people ate more than our modern dieters do. Symptoms associated with food restriction include:

Increased hunger	Nervousness	Numbness
Preoccupation with food	Secluseness	Hostility
Decreased sex drive	Aches and Pains	Polyuria
Decreased endurance	Poor concentration	Edema
Decreased activity	Poor memory	Constipation
Slow movements	Introspection	Dry skin
Decreased self-respect	Sleepiness	Dry hair
Decreased moral standards	Depression	Headaches
Poor cold tolerance	Dizziness	Apathy
Irritability	Lightheadedness	Weakness
Decreased sense of well-being	Paresthesia	Diarrhea

These symptoms are very commonly seen in those patients who have been repeated dieters. They probably result from various nutritional deficiencies or from inadequate energy intake from limited food consumption. Those who used artificially sweetened products to help restrict food intake may experience these symptoms from the starvation effect, and may or may not have additional problems from the effects of the artificial sweeteners.

We are convinced that a great number of the health problems and unpleasant symptoms suffered by many people result from not getting enough nutrients. Enough vitamins are added to our junk food to prevent formal vitamin deficiency disease like rickets or scurvy, but less complete vitamin and mineral deficiencies can certainly cause problems.

5. Toxic Effects From Aspartame

METHANOL. For every molecule of aspartame ingested, one molecule of methanol is released into the bloodstream. There is some concern about the safety of this methanol.

Methanol is a well-known toxin. Humans seem to be much more vulnerable to toxic effects from methanol than any other animal known.[79,80,81] There may be a very large variation in the amount of methanol that is toxic. It has been shown that as little as 15 mls. (1 tablespoon) of 40 percent methanol has been fatal in some cases, while other people have tolerated up to 500 mls. without permanent damage.[82,83] The usual symptoms seen in methanol ingestion start out like those of ethanol (ethyl alcohol) intoxication. Later symptoms of toxicity can include headache, dizziness, weakness, nausea, vomiting, abdominal pain, muscular pain, disorientation, visual defects (ranging from blurred vision to complete loss of vision), convulsions, and death.[84,85]

An interesting feature of methanol toxicity is the variability of time from ingestion of methanol until onset of toxic symptoms (this period is referred to as the latent period). The latent period may last only a few hours, or may last up to seventy-two hours. There's another interesting feature of methanol toxicity: the long period of time required for the effects of a single dose to disappear. Eye damage caused by methanol may last up to sixty days.[86,87] If damage to tissues occurs, eye problems can last a long time. It seems probable that even small doses of methanol ingested on a regular basis could eventually build to toxic levels.

One of the safety claims made by the producers of aspartame is that even large doses do not cause significant levels of methanol in the bloodstream. Apparently it's not the methanol itself that causes the damage, since it is gone from the bloodstream long before serious symptoms develop. Many believe that the damage results from the breakdown products of methanol (formaldehyde and formic acid), but no one is completely sure.

Several researchers have indicated that the level of methanol resulting from aspartame is far below the toxic level.[88] Some of these safety claims were based on estimates of how much aspartame people would be using.[89] What's wrong with that? Many people are far exceeding initial use estimates. It has been pointed out that the amount of methanol ingested by some people in the form of aspartame can exceed the Environmental Protection Agency's recommended limit of consumption for this cumulative toxin by thirty-two times.[90]

Some claim that the amount of methanol in diet drinks sweetened with aspartame is actually less than the methanol found in canned fruit juice.[91] Others have questioned this, showing that—in most cases—the methanol in fruit juice is significantly less than in diet drinks.[92] It is also unlikely that anyone would, or could, drink as much fruit juice as diet soda.

There are also some major differences between ingesting methanol by itself and ingesting it along with real food. Although fruit juice does break down in the can to form methanol, it also produces ethanol, which protects against methanol toxicity.[93,94] Ethanol slows the rate of conversion from methanol to formaldehyde, and allows time for some of the methanol to be eliminated from the breath and urine.[95] Most foods also contain some folic acid,[96] which is necessary for the breakdown of methanol. Under most circumstances, if you ingest methanol along with fruit juice or other food, you'd get at least two protective substances (ethanol and folic acid).

It appears that many chemicals normally found in foods protect us from various toxic substances. Dr. Stephen Levin showed that rats who were fasted for six days and then exposed to a toxic substance died at only 1/25th the dose required to kill well-fed rats. He concluded that certain antioxidant vitamins and other nutrients protected the fed rats from the toxic products.[97]

It has also been shown that a number of protective enzymes break down aflatoxins (toxic substances produced by molds sometimes found in our food supply) and other toxins. Rats who were fasted for only one day or given a highly refined diet had very little hydroxylase activity. Giving them a small amount of cabbage or brussels sprouts increased the hydroxylase activity by 100 fold, and inhibited the toxic effects of aflatoxins.[98,99]

We are constantly exposed to toxic substances from our food and drink, as well as from the fumes we breathe. By eating plenty of wholesome food (especially vegetables), we also ingest many chemicals that minimize the damage done by these toxins. Methanol that is eaten along with wholesome food—with its protective chemicals—may cause much less trouble than methanol ingested in a large amount of diet drink with no protective food chemicals accompanying it.

The amount of methanol that is toxic should depend not only on the amount of aspartame consumed, but also on one's nutritional status. Someone who drank huge amounts of diet drinks and consistently ate very little nutritious food would be more at risk from methanol toxicity than someone who drank the same amount of diet drinks along with a good, wholesome diet.

Toxic levels of methanol have been determined by giving it to experimental animals—but that's not an accurate indication! Why? Because of the unique susceptibility of humans to the toxic effects of methanol, no experimental animals provide a useful comparison. Even monkeys are not close enough to humans to reliably predict safe levels. For this reason, scientists are appallingly ignorant about the toxic effects of methanol.

Toxic levels are considered to be those which cause irreversible damage or tissue damage that can be identified with a microscope. In other words, scientists only recognize the large levels of toxins that result in tissue changes. But low levels of toxins may interfere with the function of various cells, causing unpleasant symptoms.

The repeated claims of the FDA and the producers of aspartame that it is safe do not rule out the possibility that the methanol from aspartame ingestion can lead to many unpleasant side effects.

FORMALDEHYDE. Formaldehyde is the major by-product of methanol metabolism. Formaldehyde is also present in a vast number of products to which we are all exposed on a regular basis. The air inside every modern building contains formaldehyde unless heroic measures have been taken to eliminate it. Formaldehyde is also found in—among others—the following products: clothing (wash-and-wear components, sizing, dyes, waterproofing, and anti-shrink agents), glues, adhesives, cements, paste, resins, urea-foam insulation, particle board, plywood, cellulose esters, paint, primer, paint-stripping agents, paper, wet-strength paper, polishes, waxes, disinfectants, cleansers, fumigators, cosmetics (fingernail polish and polish remover, bath oils, bubble baths, shampoos, wavesets, deodorants, and toothpaste), medication (wart remover, mouthwash, contraceptive foam, antihidrotics, denatured alcohol, cast material, dental materials, and various drugs), preservatives, preservatives in latex rubber, tanning agents, canned ice, printing/etching materials, inks, sealers, and dry cleaning spot removers.

High levels of formaldehyde exposure adversely affect everyone exposed. A number of new buildings have had such high levels of formaldehyde and other chemicals, that they could not be inhabited until exposure levels were reduced. The term **"Sick Building Syndrome"** has been given to this condition, and it has been extensively described.[100,101,102,103,104] Some of the symptoms seem to result from local contact of exposed tissues—eye, nose, and throat irritation; sensation of dry mucous membranes and skin; hoarseness, wheezing, cough, and frequent respiratory infections; itching skin; and erythema. Other symptoms are more general (they commonly include mental fatigue, headaches, hypersensitivity, nausea, and dizziness), and seem to be from formaldehyde and other chemical absorption into the blood-stream.[105]

Some people seem to be very sensitive to formaldehyde, and have similar symptoms at lower levels of exposure. There have been many reports in the recent medical literature about allergic responses to formaldehyde.[106,107,108,109]

The amount of formaldehyde from even very large amounts of aspartame may not be enough to cause problems by itself. But if you get formaldehyde from aspartame in addition to exposure from many other sources, you might get more than you can tolerate—and you might develop unpleasant symptoms as a result.

Formaldehyde dangers have been hotly debated recently. Urea foam insulation, which breaks down into formaldehyde, was tested, found safe, and approved for use in the United States and Canada. But then many people in buildings recently insulated with urea foam insulation began having medical problems, and eventually a higher-than-expected incidence of cancer was noted. The Canadian Broadcasting Company did some investigative reporting on this issue and interviewed some experts, who gave assurances that urea foam had been adequately tested, that it was perfectly safe, and that it caused no problems.[110] (This reassurance sounded a lot like the reassurance given by aspartame researchers and the FDA). Since that time, however, researchers concluded that the foam did cause a health hazard; it was banned from further use, and millions of dollars have been spent removing the foam from many buildings. In recent research, formaldehyde has clearly been shown to cause cancerous and pre-cancerous lesions.[111,112,113]

ASPARTIC ACID. By weight, about 40 percent of aspartame breaks down to the amino acid aspartic acid, also called aspartate. The concentration of aspartic acid and a similar amino acid, glutamic acid, is particularly high in the brain, and may make up 25 to 30 percent of the total free amino acid pool.[114] Aspartic acid is an excitatory neurotransmitter substance,[115] which in high doses has been shown to kill brain cells in experimental animals.[116,117,118,119,120,121]

Very little work has been done on aspartic acid by itself. Its close relative, glutamic acid (glutamate), however, has been widely studied in the form of monosodium glutamate (MSG). These two amino acids are similar in structure, are transported by the same amino acid carrier system, and have been shown to have very similar toxic effects. The following discussion of glutamic acid, then, should apply equally to aspartic acid.

Glutamic acid (like aspartic acid) has been shown to kill brain cells in a wide variety of experimental animals. A review article written eight years ago described more than sixty studies that had demonstrated brain damage from various levels of glutamic acid.[122] Glutamic acid has also been shown to cause behavior and learning changes in animals.[123,124,125] Changes in various hormones and in the female reproductive system have also been caused by glutamic acid.[126,127,128]

Humans respond to aspartic acid and glutamic acid in a unique way when compared to experimental animals. Figure 8 compares the serum levels of glutamic acid after administering equal amounts by weight to monkeys, mice, and humans.[129] As is shown, serum levels of glutamic acid in humans quickly rise to much higher levels than in mice or monkeys. (Note that monkeys are not always a good experimental model to predict human response to an experimental situation).

When aspartic acid is ingested with protein or carbohydrates, the rise in blood levels is much less than when aspartic acid is ingested by itself.[130] Note that in Figures 9 and 10, the levels of glutamic acid are considerably higher when free glutamic acid is given, compared to when given with carbohydrate (CHO) or in a form bound to protein. In other words, if you get aspartame in diet pop with no other nutrients, you'll get a higher level of serum aspartic acid than if you ate aspartame along with food.

It has been suggested that the levels of aspartic acid in the human bloodstream, even after very large amounts of aspartame, are not high enough to cause any brain damage in humans.[131] Whether aspartic acid levels are high enough to cause symptoms is not known at this time.

PHENYLALANINE. The amino acid phenylalanine is essential for normal brain development and function, but in excessive amounts is well known to cause brain damage, severe mental retardation (with I.Q. as low as 20), aggressive and hyperactive

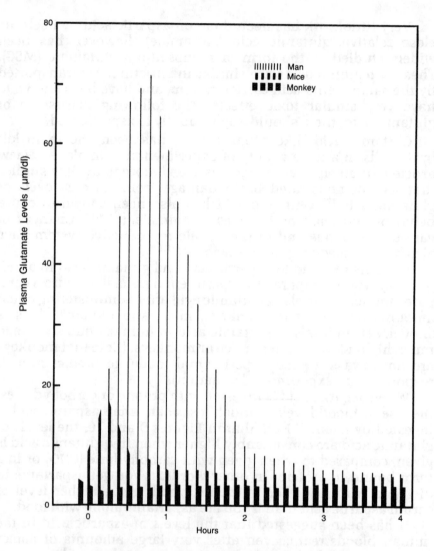

Figure 8. Mean glutamic acid (glutamate) level in adult humans, mice, and monkeys given 150 mg MSG/kg body weight dissolved in water. Similar results would be expected with aspartic acid. Drawn from data presented by Stegink, et al.[129]

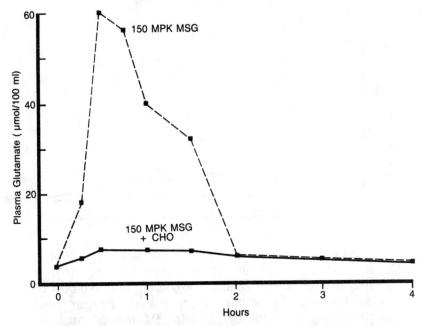

Figure 9. Mean plasma glutamic acid (glutamate) concentrations in normal adult subjects ingesting 150 mg/kg body weight (MPK) monosodium L-glutamate (MSG) with and without 1.1 g/kg carbohydrate (CHO) in the form of polycose. Similar results would be expected from aspartic acid. Drawn from data presented by Stegink.[130]

behavior, and seizures.[131] Phenylketonuria (PKU) is an inherited enzyme deficiency that leads to excessive levels of phenylalanine, and has been extensively studied. Since aspartame boosts phenylalanine levels in the blood and the brain, studying PKU can be a good way to understand what happens as a result.

Although only 1 in 15,000 babies is born with PKU,[133] 2 percent[134] to 10 percent of our population[135] may be carriers for the genetic trait. Those with PKU lack the enzyme necessary to break down phenylalanine, and those who carry the genetic trait have about 10 percent of the normal levels of this enzyme.[136] Under normal circumstances, the carriers have no problems—but those

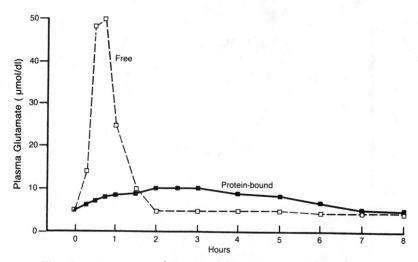

Figure 10. Mean plasma glutamic acid (glutamate) in normal adults ingesting either 100 mg of free glutamate per kilogram body weight dissolved in 4.2 ml/kg body weight water, or a hamburger meal providing 1 gram of protein, 94 mg of protein-bound glutamate, and 68 mg of protein-bound glutamine per kilogram body weight. Drawn from data presented by Stegink.[130]

with PKU will develop brain damage unless they are put on a special diet to keep the phenylalanine levels down.

Autopsies of the brains of PKU patients showed a five-fold increase in phenylalanine and decreases in tyrosine, tryptophan, serotonin, dopamine, and norepinephrine.[137]

Some experts have blamed the high phenylalanine or its metabolic by-products for the brain damage in PKU.[138] Others have blamed the low levels of the other neurotransmitters.[139,140]

It is clear that large amounts of aspartame can cause an increase in the phenylalanine levels in the blood and brain tissue of normal people.[141,142,143,144] It is also clear that aspartame can cause the other neurochemical changes characteristic of PKU. The major difference is the magnitude—the changes in PKU are more extreme than the changes from ingesting aspartame. But several situations may cause levels of phenylalanine to become higher than normal after ingestion of aspartame.

The ingestion of aspartame along with carbohydrate will cause higher blood levels of phenylalanine than when aspartame is ingested by itself.[145] Remember that both aspartic acid and methanol (the other components of aspartame) are more toxic and produce higher blood levels when aspartame is ingested on an empty stomach.

Carriers of the PKU trait develop levels of phenylalanine in the serum approximately twice that of normal people after aspartame ingestion,[146] and take about twice as long to clear it from their systems.[147]

Those with liver disease[148] and kidney disease[149] do not clear phenylalanine from their bodies as quickly as normal people. Estrogen supplementation may also cause an impairment in phenylalanine metabolism.[150]

Pregnancy can slow down the rate of phenylalanine metabolism.[151] In early pregnancy (when the most vital fetal developments occur), phenylalanine levels in pregnant rabbits given aspartame can become three times higher than in later pregnancy.[152] Phenylalanine levels can become concentrated on the fetal side of the placenta, and blood levels there can become twice as high as on the mother's side.[153,154] A fetus is also less capable than the mother of metabolizing phenylalanine.[155] This combination of problems during pregnancy may especially be a problem for carriers of PKU, and some think that fetal brain damage may result from aspartame in pregnancies among this group of women.[156,157]

Here's the important question: are these neurochemical changes enough to cause any danger or side effects? Some experts have suggested that even with abuse doses of aspartame, phenylalanine levels don't get high enough to be toxic. They maintain that the increased levels of phenylalanine from aspartame are well below the level at which brain damage occurs.[158]

There is a conflict about how much phenylalanine in the blood can cause brain damage. Some have suggested that there is a **threshold level—above it, damage occurs, but below it, no problems arise.** Others have suggested that progressively more damage occurs as phenylalanine levels rise.[159] The level of serum phenylalanine for a pregnant woman above which retardation occurs in the baby is .6 to 1 mM/l.[160,161] It has been shown, however, that levels as low as .1 to .3mM/l may cause a ten-point drop in I.Q. in the baby.[162,163] There is also an impairment in brain function (neuropsychologic performance) at these lower levels, suggesting that more damage occurs as phenylalanine levels rise.[164]

If levels of phenylalanine much lower than that causing profound mental retardation can cause permanent damage with reduced I.Q., then we would certainly expect subtle effects on the brain (with altered function and unpleasant symptoms) at even lower levels.

Phenylalanine also seems to have an impact on obesity. Obese patients have higher blood levels of phenylalanine and a few other amino acids than do thin people. These elevated amino acid levels may be partly responsible for the higher-than-normal insulin levels found in the obese.[165] Several amino acids (including phenylalanine) stimulate insulin production. When given along with carbohydrate, phenylalanine causes a much higher increase in insulin than sugar by itself.[166] (See Fig. 11.) The sudden increase in phenylalanine in the blood following ingestion of aspartame may contribute to weight gain through this effect on insulin. This tendency to cause weight gain would be even greater if carbohydrate were consumed along with the aspartame.

Experimental Studies for Aspartame Safety.

A number of studies have been performed using various doses of aspartame on different species of experimental animals to assess safety for humans. Some of these studies were done with newborn animals, others with older animals. A number of abnormalities were noted. In mice, there were abnormalities in various blood and clinical chemistry tests.[167] Male rats had abnormally small hearts, and female rats had abnormally large livers. Female rats also had an increased incidence of liver hyperplastic nodules.[168] An increase in relative spleen and kidney size was observed. Deposits of calcium were found in the kidney, which reflected increased loss of calcium through the kidneys in these animals.[169] Deposits of hemosiderin, focal hyperplasia, and tubular degeneration were also observed within the kidney tissue of male rats. The death rate was higher in one group of female rats given aspartame.[170]

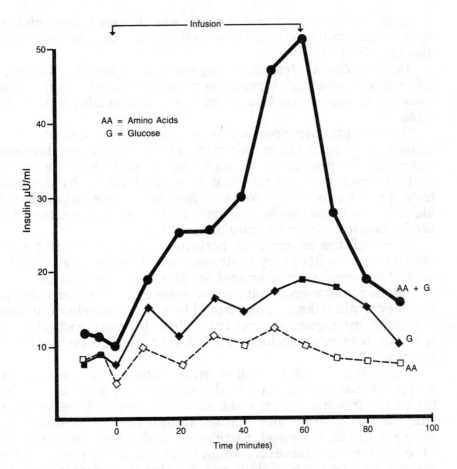

Figure 11. Effect of amino acids (AA), glucose (G), and a combination (AA+G) on plasma insulin. An infusion of a 27 gram mixture of the amino acids phenylalanine, leucine, valine, and isoleucine was given over a 60 minute interval, with and without a glucose infusion of 100 mg/minute. Note that the combined infusion had a much greater effect on insulin secretion than glucose by itself. Drawn from data presented by Ogundipe and Bray. (The Obese Patient - G. Bray.)

Dogs suffered impaired growth and lowered hemoglobin, hematocrit, and total red blood cell count. Liver function was also impaired.[171]

Developmental defects,[172] impairment in learning (delayed reflex development, delayed swimming skills, and so on),[173] and behavior changes were also noted in young animals given aspartame.[174]

Some of the safety reassurances in human studies were based on blood values taken on a fasting basis, when peak levels of aspartame by-products may have skyrocketed shortly after ingesting aspartame but returned to normal values by the time testing was done.[175,176,177] None of these tests rules out the possible side effects that could arise from toxic levels of aspartame by-products interfering with cellular function.

Some of the researchers performing animal studies with aspartame have been very cautious about drawing safety conclusions for humans from animal studies. One scientist said, "It should be remembered that cross-species generalizations always involve certain risks . . . one would be wise to keep in mind that stumptail macaques are not furry little humans with stumpy tails but, instead, members of a different (albeit closely related) species."[178]

Regarding human studies with aspartame, another investigator has stated, "All of the studies can be criticized for their experimental design and because the results have not been subjected to sophisticated statistical analyses. Numerous questions remain unanswered, particularly with respect to the possible effects of aspartame on human beings who may consume the compound over years of daily use."[179] One investigator expressed concern that in some vulnerable people, "abuse levels of intake . . . might approach a zone where fetal toxicity would be possible."[180] One researcher concluded that ". . . data from preclinical studies . . . cannot by themselves assure that a food additive is completely safe for all segments of the consuming population. . . . Only use, concomitant with regulatory surveillance, can lead to a final determination of whether aspartame is safe for use by all segments of the population."[181]

Let's get back to the original question: are artificial sweeteners safe?

More than ten thousand people have reported sides effects after using aspartame.[182] Although not completely proven, it appears almost without question that aspartame can cause a wide range of symptoms. That can hardly be considered safe.

Other food additives that were originally thought to be safe have been proven to have a great number of side effects, and there are even more possible dangers under certain circumstances. It is also likely that aspartame will eventually become identified with specific problems as more experience is gained with it, and more research is done.

Until that research is done, it's up to you—are you willing to take the chance?

CHAPTER SIX
THE SUGAR ISSUE

Chances are, you've heard plenty about sugar.

And, chances are, what you have heard is full of confusion and conflict.

Why?

The role of refined sugar (sucrose) in human nutrition is an extremely controversial topic, that's why. Opinions cover the entire spectrum: some believe that any sugar is harmful and that it should be entirely eliminated from our diets, while others believe that it is nutritious and should be increased in the diets of those suffering from food shortages and malnourishment.[1]

Since artificial sweeteners affect the amount of sugar you eat, we need to take a close look at sugar. In some cases, people actually do cut down the amount of sugar they eat by using artificial sweeteners as an alternative, but in most cases artificial sweeteners actually increase the desire for and the consumption of sugar.

Over the last fifteen years, we have personally worked with about six thousand people who, for various reasons, were encouraged to completely eliminate sucrose, other refined sugars, chocolate, and caffeine from their diets. In the last few years, we have also asked them to eliminate artificial sweeteners.

What have we learned?

We'd like to share some of what we've found in working with these patients. About four thousand of them were on a very low-calorie diet that did not provide enough essential nutrients. (This was back in the days when low-calorie diets were thought to be the only way to lose weight, and before we developed our current weight-loss methods). In the last approximately two thousand patients, the no-sugar regime was used along with a program of eating enough nutritious, low-fat food on a regular basis to completely satisfy hunger.

There were some similarities between the two groups:

Most patients suffered **withdrawal symptoms;** these withdrawal symptoms were part of an **addiction** to the sugar, always included a strong desire for sugar-containing food, and often included headaches, irritability, anxiety, sleep disturbance, body aches, fatigue, and depression. In most cases, withdrawal symptoms were reasonably tolerable, and usually disappeared in four to ten days.

In some cases, withdrawal symptoms were very severe. One lady in her late thirties who stopped eating sugar completely found that her cravings diminished within a few days. But other symptoms intensified. Headaches were severe. She got progressively weaker and more depressed. She began spending most of her time in bed, not caring whether she lived or died. After three weeks, she finally gave up, ate a doughnut, and within minutes felt rejuvenated. Her headaches and depression lifted, her energy began to return, she got out of bed, got dressed, and went to work. She eventually was able to stop eating sugar completely, and then felt considerably better in many ways than she did while using it.

We have tried to figure out how common sugar addiction is in our society. From questioning our patients and from taking informal polls at seminars and speaking engagements, we believe that at least three-fourths of the patients we see for weight control are addicted to sugar. But we realize that we see a rather select group of people, and the overall incidence of sugar addiction in the population may be less than this.

Headaches often ease up or disappear when people stop eating sugar. This improvement is seen whether the headaches are migraine, tension, sinus, or other unspecified headaches. Although the headaches very frequently intensify for the first few days, they eventually go away completely in most cases. If headaches intensify at first, they are clearly withdrawal headaches from addiction to the sugar. Some people also get headaches soon after ingesting sugar. Sometimes people get one kind of headache from eating sugar, and another type when going through withdrawal. When headaches occur both from ingesting sugar and withdrawal, they are usually daily, and become constant or almost constant. The only relief occurs during the few hours after ingesting sugar when the acute effects are gone, and before the withdrawal headaches begin.

Sleep disturbances often improve after eliminating sugar. Trouble getting to sleep, trouble waking up, and an inability to get back to sleep are common in our population—but, in many cases, they improve within a few days of sugar avoidance.

When people stop eating sugar, **depression** often improves. **Anxiety and irritability** commonly improve as well. Many patients have been troubled with both depression and anxiety, and frequently there are wide swings in mood from one time of the day to another. These mood swings also improve with sugar avoidance.

Energy levels often improve dramatically from sugar avoidance once the withdrawal phase is over. Sugar users tend to have very erratic energy levels, with surges of energy after sugar ingestion followed within a few hours by very low energy levels. Energy may go up and down like a yo-yo, and these people often begin to use sugar products (especially those containing caffeine) to give themselves energy. When the sugar habit is finally kicked, they notice a much smoother energy level, with very little change from one time of the day to the next.

Arthritis, muscle aches and pains, indigestion, gas, bloating, ulcer pains, heartburn, various other digestive problems, and a wide range of other symptoms will often disappear as well when people cut out sugar. Most people with sugar addiction have used it daily for many years, and had no idea that it could cause any of these problems.

One of the intriguing things about these sugar-induced problems is that they often *immediately* occur if you eat sugar again. Someone who has been off sugar for several months might indulge once more at a party—only to suffer a return of the old symptoms within a few minutes. And the symptoms aren't always fleeting—the headache, arthritic pain, or other symptoms may become very severe, and may last for several days.

The effects described above have obviously been related to sugar ingestion, and were seen in both the starvation dieting group and in the group eating plenty of food. There are some very major differences in these two groups, however. Although some of the people in the starvation group experienced improved energy from the avoidance of sugar, virtually all of this group experienced a gradual decrease in energy, became weaker, found their endurance decreasing, and had less desire to do active things. Even though many of them felt better emotionally from the sugar avoidance, most of these people eventually became more depressed, apathetic, and irritable as time went on, and began to have trouble sleeping.

The experience of those in the group who were eating plenty was dramatically different. Although the same things happened at first to those in both groups, the people who ate adequately noticed a steady improvement in their strength, energy, and endurance. When the emotional problems were related only to the sugar intake, they improved with sugar avoidance alone. Most overweight people who have tried dieting—and many who for various other reasons were undernourished—experienced some degree of depression, anxiety, and irritability. These problems gradually improve within several weeks to several months with continued good nutrition.

The dramatic improvement in so many symptoms simply with sugar avoidance and good nutrition has convinced us that a high percentage of the medical complaints that we commonly see in our population are simply products of poor nutrition. We as a society seem to have lost track of the very basic concepts that if your brain and body are going to function properly, they have to be fed properly.

Many people know that they do not feel as well when they eat sugars, but do sugars have any bearing on actual diseases and the risk of dying?

Medical problems aggravated by refined sugar.

The role of sucrose in relationship to various disease processes has been extensively investigated. Let's look at the results of some of those studies, which used animals as well as humans:

Sugar Metabolism and Diabetes. In experiments in which animals are given various amounts of sucrose, a number of studies have shown increased levels of insulin.[2,3,4] Other studies showed an increase in levels of both insulin and glucose in the bloodstream after eating sucrose.[5,6] A number of studies showed that sucrose impaired glucose tolerance or caused insulin resistance (basically the same thing).[7,8,9,10]

Similar results have been shown in humans. Some studies have shown increased glucose, insulin, or both after eating sugar.[11,12,13] Glucose tolerance was also impaired in humans.[14]

In diabetic animals[15,16,17,18,19] or diabetic humans,[20,21,22] sucrose intake had an even more profound impact on blood levels of insulin, glucose, and on insulin resistance. In humans, the effects of adding extra sucrose are not quite as dramatic as they could be, since almost everyone in our society is already using quite a bit of sugar. One good method of evaluating the effect of sucrose on diabetes is to strictly avoid sucrose, and then measure various components of sugar metabolism. This was done, along with increasing the complex carbohydrate and fiber.[23,24,25,26] In these studies, the insulin resistance decreased, the insulin requirements decreased, and, in many of the subjects, the insulin dependence disappeared (they were able to stop using insulin and control their diabetes with diet alone)!

As pointed out in Chapter Four, another way to evaluate the effect of sugar is to compare the people in countries where little sugar is eaten with those in countries where a lot of sugar is eaten in the average diet. Typical of these findings is the observation that Australian Aborigines who adapt to an urban lifestyle (which includes sucrose) develop a high incidence of diabetes, even though they have a very low diabetic incidence with their native diet (which does not include sucrose).[27]

The worsening of diabetes, insulin resistance, and higher levels of insulin and glucose with the addition of sucrose to the diet leaves little doubt that sucrose has a negative influence on diabetes. Further proof comes from the improvement of diabetes with a sucrose-free diet, and the low incidence of diabetes in populations with very low sucrose intakes.[28]

Sucrose Effects on Obesity. Many studies have clearly shown that eating sugar helps make people fat.[29,30,31,32,33,34,35,36,37,38,39,40] In some studies even though weight increased only slightly, body fat actually doubled! Other studies have shown a considerable increase in body weight when sugar is added to the diet.[41] One study showed a ten-fold increase in lipid synthesis in animals given sucrose.[42] Increased body weight has been demonstrated by adding sucrose to the diets of rats, even though they didn't eat any more calories than a control group that ate starch.[43] Fat gain has even been demonstrated in animals given sucrose in a situation in which they ate fewer calories (their food contained more than 80 percent sucrose, which is not very appetizing).[44]

Different experimental animals seem to respond somewhat differently to sugar, and different strains of the same species also respond differently. Some strains are much more likely than others to gain weight when their diets contain sugar. Humans also vary considerably, and some people are much more prone to gain weight and body fat when they eat sugar. Although "normal" people have been shown to gain weight when given diets high in sugar,[45,46,47] those people with an abnormal glucose tolerance seem to gain weight more easily when given extra sucrose.[48]

The combination of sucrose and extra fats cause more weight gain than sucrose itself in animals[49,50,51,52] as well as humans.[53,54]

High Blood Pressure, Heart Disease, Strokes, and Elevated Lipids. Sucrose has been shown to cause elevated blood pressure in experimental animals[55,56,57,58,59] and humans.[60,61] The blood pressure elevation is higher in those subjects who also ate salt.[62,63,64] Animals eating sucrose have been shown to develop higher serum triglycerides[65,66,67,68,69,70,71,72,73] and, in some cases, increased cholesterol[74,75] or a combination of increased triglycerides and cholesterol.[76,77,78,79,80]

An increase in the blood pressure, cholesterol, or triglycerides are all risk factors for developing arteriosclerosis (hardening of the arteries), which can lead to stroke and heart disease. Besides causing these risk factors, sucrose has also been shown to also cause arteriosclerosis.[81,82]

Cancer. A few studies have shown that human populations consuming higher amounts of sucrose also have higher incidence of some kinds of cancer.[83,84] It is quite clear from looking at the cancer statistics from one country to another that the countries whose people eat very little sugar have a very low incidence of most types of cancer, compared to countries where diets are high in sugars.[85] Since people in the high-sucrose countries also consume other refined foods and higher levels of fat as well, the link between sucrose and cancer is not clear.

Problems with sucrose research.

Obtaining information from sucrose research is a lot harder than you might imagine. If you add extra sucrose to the average diet, you also add a great number of extra calories, which may change many metabolic hormones and which will cause weight gain in most cases (both because of the nature of sucrose and from the increased calories). If you substitute sucrose for other components in the diet, then you also change the test results. If you substitute sucrose for fats, and less fat is eaten as a result, you may actually see a lowering of cholesterol levels and a decrease in weight, since fats affect these two factors more than sugar does. If you substitute sucrose for unprocessed food containing mainly complex carbohydrates with lots of fiber, vitamins, and minerals, then you would be expected to see changes made as result of deficiencies of these nutrients.

Let's take a look at some of the reasons why eating sugar can result in problems.

Reasons for sugar-related problems.

Addiction clearly accounts for many of the problems related to sugar. Addiction to sugar or food in general is not well-recognized by scientific researchers. A recent computer search of the world literature produced only one article dealing with food addiction. A number of books dealing with food allergy, however, have discussed this issue.[86,87,88,89,90]

Allergy may account for many of the problems observed with sugar ingestion. The classical IgE allergy reaction to sugar is probably uncommon, but allergy reaction in the more general sense may be relatively common. Sometimes reactions can be rather severe. A patient who had been completely off sugar for more than six months described a recent reaction to sugar in frozen yogurt. She was able to tolerate fructose, but not sucrose, and thought this particular yogurt contained fructose. Within a few minutes of eating it she began to feel dizzy and hot. A terrible headache began, and she felt profoundly fatigued. Within an hour, severe diarrhea began, and it lasted several hours. It took her two days to completely recover from this episode.

It is quite common for someone who has stopped using sugar to notice a decrease in nasal congestion and post-nasal drip. This may represent an allergic reaction to the sugar. Some of the improvement in emotions and energy may also represent a simple adverse or allergy reaction to the sugar.

Allergies likely occur to the cane or beet impurities left within the processed sucrose. Corn is well recognized to be allergenic, and it is quite common for people to react to corn-derived sweeteners (dextrose or high-fructose corn syrup). Corn-derived fructose is sweeter than sucrose, and can be less expensive for the manufacturer to provide the appropriate sweetness level, so it has become more popular in recent years. (You may recall all the publicity surrounding the changing of Coke's® formula to include corn fructose instead of sucrose. There was a lot of rebellion among Coke® fans, and eventually the original formula was reintroduced.)

Vitamin and Mineral Deficiencies may occur with excessive use of sucrose. Since sucrose has a lot of calories and absolutely no vitamins or minerals, it may provide the needed energy, but not the other nutrients. Continued use of sucrose in moderate to high levels may cause a deficiency in a number of essential nutrients. This seems to be even worse in those overweight people who, through repeated dieting, have trained their bodies to get by with a relatively small food intake. The regular diet of every overweight sugar-user that we have analyzed with our computer has shown deficiencies in at least one nutrient, and usually in many of the essential vitamins and minerals.

Although we just don't see any of the classical vitamin deficiencies like rickets and scurvy, it seems obvious that many of the common symptoms we see in our patients result from nutritional deficiencies. In many cases they gradually disappear with nothing but nutritious eating.

Mineral deficiency in particular seems to be a major problem associated with high sucrose intake. In fact, one way that experimenters produce an animal model with a mineral deficiency state is to give sucrose at high levels in place of unprocessed food that contains minerals. Deterioration in sugar metabolism and insulin resistance has been shown with refined food intake from producing deficiencies of copper,[91,92] magnesium,[93] zinc,[94] chromium, and selenium.[95] Copper deficiency has been reported in as much as 70 percent of the population in the United States, and is made worse through ingesting refined sugars. A study with rats who were copper-deficient and who were fed a high-fructose diet showed an interesting pattern of heart disease. The male rats looked and acted completely well, and then suddenly died of heart disease.[96] This is very similar to men within our population who eat a poor diet, but seem to be completely healthy until they suddenly die of heart disease.

In populations consuming lower levels of potassium, calcium, and magnesium, a higher incidence of blood pressure has been shown.[97,98,99,100] Supplementing these minerals has also been shown to decrease blood pressure.[101,102,103,104,105]

Even modest decreases in some of these minerals can make significant differences in health. A long-term population study in this country compared a group of people who died of stroke to those who didn't. Researchers concluded that if the people who died had simply eaten the amount of extra potassium found in one serving of vegetables or fresh fruit daily, they would have reduced their risk of dying by 40 percent.[106]

There is a lot of concern today over osteoporosis, which occurs in epidemic proportions in our female population. Some believe that many of these vulnerable women are not eating enough calcium, and calcium supplements have often been suggested. Diets high in sucrose and other refined foods may be a factor in producing an inadequate calcium intake. Even worse, sucrose has been shown to cause an increase in calcium and magnesium excretion from the kidneys.[107] If sucrose contributes to both an inadequate intake and a loss of the calcium and magnesium from the limited stores, it could be a major factor in causing osteoporosis.

The Balance of Intestinal Microorganisms is affected to a great extent by the diet. Meat eaters have quite a different balance of bacteria and other organisms in the intestinal tract than do strict vegetarians.[108] Sugars may selectively feed some microorganisms more than others. Intestinal yeast (Candida) seem especially to flourish with refined sugar. If the repeated use of antibiotics have eliminated many of the bacteria that compete with yeast, they may become plentiful enough to cause problems in the presence of refined sugars. They are well known to cause vaginitis and thrush (yeast infection of the mouth, usually in babies). They can also cause skin infections—especially diaper rash in babies—and rashes in moist areas of diabetics and others with poor immunity. They can invade the intestinal tract, leading to abdominal pain, gas, bloating, diarrhea, constipation, heartburn, and a wide range of other symptoms.

Yeast produce chemical by-products from their metabolic processes, which in large amounts can be rather toxic. These toxins can interfere with various enzyme systems, and cause a wide range of symptoms throughout the body. Brain function seems particularly impaired, and typically, yeast can cause trouble with thinking clearly, memory loss, irritability, fears, anxiety, depression, headaches, trouble focusing, and dizziness. These toxins seem to interfere with the enzymes that control sugar metabolism, and cause a lot of symptoms similar to those caused by low blood sugar—such as excessive hunger, weakness, dizziness, headaches, and other symptoms a few hours after eating. Yeast overgrowth seems to cause a lot of muscle aches, pains, stiffness, and weakness.

When people who have typical yeast-related symptoms eat food containing a lot of refined sugar, they often notice a dramatic increase in symptoms within as little as fifteen to sixty minutes. This is even more dramatic if they had been completely off sugars for a few weeks prior to indulging. Some people have described the gradual onset of fatigue, generalized aching, mental dullness, depression, headache, and a feeling like they have been poisoned. Symptoms like these that clearly occur after eating sugars are very suggestive of yeast overgrowth.

Any plant that survives in the soil contains a number of natural fungus inhibitors that allow the plant to survive the molds and fungi which would otherwise kill and break them down.[109,110,111] All plant materials that are unprocessed and used for food contain these chemicals. They can inhibit yeast overgrowth[112,113,114] while the refined sugars can cause rapid intestinal yeast growth. This rapid multiplication and activity of intestinal yeast appears to be much like the activity of bread yeast when it is fed sugar.

The activity of intestinal yeast with its toxic by-products may be the reasons many people don't feel as good after eating food with a lot of sugar compared to the way they feel when they eat unprocessed food. Even people who have no symptoms unless they eat sugar could be suffering from yeast overgrowth.

Yeast overgrowth appears to be very common, and we have now treated about 2000 people suspected of having this problem. We have studied this problem very extensively, and gathered about 3,000 reference articles related to yeast in the process of writing our book, **Back To Health: A Comprehensive Medical and Nutritional Yeast Control Program** (Vitality House International, Inc. 1986). A number of other books have also been written about this problem in the last few years.[115,116,117,118] In spite of the thousands of doctors who are recognizing and treating yeast overgrowth, many doctors don't yet believe that yeast can cause any of these problems.

Low Blood Sugar (Hypoglycemia) may occur a few hours after eating refined sugar products. This may also occur after eating some other carbohydrate sources that can be broken down quickly to form sugar.[119] Refined products may cause worse problems over the long term, however, because of the lack of minerals that makes the sugar metabolism progressively worse. Hypoglycemia probably doesn't occur unless a number of problems exist within the system responsible for normalizing blood sugar levels. Unprocessed foods that take a long time to break down and that release sugar and other nutrients into the bloodstream gradually over a period of hours are less likely to cause this problem.

What do the researchers say about sugar?

A great deal of research has now been done on sucrose and other refined sugars by many different researchers. Much of this research has been funded by producers of sugar and sugar-containing products. A very comprehensive review of the scientific literature has recently been undertaken by a special committee appointed by a branch of the FDA. Committee members reviewed approximately one thousand articles, and reported their findings in a "**Report From FDA's Sugars Task Force: Evaluation of Health Aspects of Sugars Contained in Carbohydrate Sweeteners.**"[120]

How did the committee handle all the conflicting information on sugar?

Regarding the issue of whether or not sucrose causes hypertension, they pointed out many studies in which the subjects using sugar did have elevated blood pressure compared to a control group, and some studies in which they did not. They suggested that the observed increase in blood pressure was caused by weight gain, and then proceeded to deny that both weight gain and blood pressure elevation occurred as a result of eating sugar. Specifically, they said that ". . . the body weight also increased during the experimental period, supporting the possibility that the observed effect on blood pressure may be related to the body weight increase and not specifically to sucrose per se." "No evidence was found to support the contention that **current dietary intake** of sugars contributes to the development of hypertension.", and that ". . . sugars do not have a **unique role** in the etiology [cause] of obesity."

Most of the conclusions of the Sugars Task Force used scientific hedges. You may be interested in some of their main conclusions:

"Although consuming diets with very high levels of sugars may produce adverse effects on glucose tolerance and insulin metabolism, there is **no persuasive scientific evidence** that sugars as they are **currently consumed** by the U.S. population are an **independent risk factor** associated with the development of impaired glucose tolerance. . . . **present levels** in sugars consumption patterns are **relatively insignificant factors** in the management of diabetes mellitus." "There is no **conclusive evidence** that dietary sugars are an **independent risk factor** for coronary artery disease in the **general population**." "There is no **firm evidence** that sugars as **currently consumed** interfere with the bioavailability of vitamins, minerals or trace nutrients . . ." "There is no **convincing scientific evidence** that sugars consumption is an **independent risk factor** in the production of gallstones." "There is no **substantive evidence** that the consumption of sugars is responsible for behavioral changes in children or in adults **with the exception** of the relatively rare hypoglycemias that are present in the population."[121]

The final conclusion of the Sugars Task Force?

"Other than the contribution to dental caries, there is no **conclusive evidence** on sugars that demonstrates a hazard to the **general public** when sugars are consumed at the levels that are now current and in the manner now practiced."

The conclusions that these scientists reach in the face of almost overwhelming evidence that sucrose causes health problems seem incredulous. Many scientists will continue to support traditional concepts until the evidence against those concepts becomes overwhelming. The effects of refined sugars on health is still a controversial issue, and it may take many years before the truth is apparent. These task force scientists appear to be saying that sugar consumption at its present level in our society is ideal.

Sugar consumption in our society is not ideal. We have a tragically high incidence of a wide range of serious illnesses that seem preventable simply with a healthier diet. The evidence that sugar contributes to these diseases is strong enough that anyone interested in better health would do well to heed the advice of several other groups of scientists who have repeatedly encouraged people to reduce their intake of refined sugars and increase their intake of fiber and complex carbohydrates (see Chapter Three). The harmful effects of sugar are intensified by cholesterol and other fats in the diet.[122]

Normal-weight people who eat plenty of nutritious, high-fiber, high-complex-carbohydrate, low-fat food can probably eat small to moderate amounts of refined sugar without discomfort or significant health problems—if they don't have an adverse reaction to the sugar and don't develop an addiction to it.

What about the rest of us?

Anyone who uses enough sugar to reduce their protective nutrient intake can expect a wide range of health problems that can not only make them uncomfortable, but that can increase their chances of premature death. If you have high blood pressure, elevated cholesterol or triglycerides, heart disease of any kind, diabetes or an altered glucose tolerance, kidney stones, osteoporosis, overweight, or a tendency towards any of these conditions, you'd be wise to cut way back on sugar.

If you have headaches, depression, anxiety, mood swings, sleep disturbance, fatigue, arthritis, and intestinal disturbances, you might do well to steer completely clear of refined sugar!

SECTION I - CONCLUSION

Proponents of artificial sweeteners often use the following logic to promote their products:

1. Obesity (excessive fat stores) is the result of overeating.

2. Obesity and the excess calories and carbohydrates associated with it cause serious medical problems, such as hypertension, heart disease, and diabetes.

3. Reducing calories and carbohydrates will help control weight.

4. Reducing calories and carbohydrates will reduce the risks from these obesity-related disorders.

5. Artificial sweeteners are useful as a substitute for sugar, and allow the pleasure of sweet-tasting food and drinks while helping to reduce carbohydrates and caloric intake.

One of these researchers concluded that "the cancer risk of the carbohydrates that saccharin replaces is several hundred times greater than the cancer risk from saccharin."[1] A widely publicized statement by the Calorie Control Council said that "just replacing 10 billion cans of [diet] soft drink with regular soft drink will lead to an extra two trillion calories, and that means 600 million extra pounds of fat in Americans...that could mean maybe as many as 25,000 extra heart attacks."[2]

Although these statements and the reasoning behind them seemed logical and supportable a number of years ago, it is now very obvious that both the logic and the conclusions are wrong. Let's look at each of the points individually.

The first point is obviously incorrect. Many studies have shown that the average overweight person eats less, or at least no more than, the average thin person. Obesity appears to be a problem of **fat store regulation** rather than a problem of simple energy imbalance. *Overweight people have a number of metabolic differences from thin people.* At least some of these differences seem to be caused by deficiencies of various nutrients, which result from restrictive dieting.

Now for the second point. It is true that obesity is **associated** with various medical problems including hypertension, heart disease, and diabetes. In other words, obese people do have a higher incidence of these problems than thin people, according to some research. Since overeating does not cause obesity, then overeating clearly cannot cause these obesity-related disorders. These problems are then aggravated by either the excess weight itself, or possibly by the decreased nutrients consumed by the obese. Since obesity and eating less are so inseparably connected, it is hard to know whether the weight itself is a factor. It is clear, however, that eating less is a definite risk factor for developing or dying from these disorders. Study after study on thousands of people have repeatedly confirmed this finding. Reduced-calorie dieting, then, is a definite contributor to various illnesses, and appears to be even more dangerous than the increased weight.

Let's look at the third point. It has been clearly established that cutting down on caloric intake and carbohydrates will cause short-term weight loss. In spite of continued efforts to restrict food intake, however, very few people will keep weight off when it is lost with reduced-calorie dieting. Close to 100 per cent of those who attempt weight loss in this way will eventually fail. Failure means, in many cases, gaining more weight than ever. Dieting then, makes you fatter. Reduced- calorie dieting not only fails to correct any of the metabolic problems associated with obesity, but seems instead to make them worse.

Now for point four. In view of the extensive evidence that eating more food provides some protection against developing hypertension, strokes, heart disease, and cancer, it goes against all reason to deliberately cut calories. Since carbohydrates, fiber, and various vitamins, minerals, and other chemical substances contained in high-carbohydrate foods have all been shown to be protective, it also makes no sense at all to reduce carbohydrates.

Now to point five. To replace some of the refined sugars in the diet is certainly a worthwhile objective. Many of our patients report not feeling as well while regularly eating sugar-containing foods. The vast majority of our overweight patients report clear-cut addictions to sugar. Sugar intake seems to contribute to high cholesterol and other lipids in the blood stream, hypertension, heart disease, and even osteoporosis. It is not clear whether there is anything about sugar itself which causes these problems, or whether it is just the fact that refined sugar is used in place of nutrient-containing foods which would otherwise protect against developing these conditions. The role of refined sugars in causing obesity is very clear. It seems to increase the efficiency with which fat is deposited, and can lead to considerable weight gain even with no added calories.

If artificial sweeteners helped replace sugar in the diet, then they would be very useful indeed. We do concede that most people can reduce their sugar and total food intake on a **short-term basis** using artificially sweetened products **if they use them along with a conscientious effort to follow a formal, calorie-restricted diet.** There are even some people who can use diet drinks to entirely replace breakfast or even lunch. A few people can even sustain a low caloric intake for a long term with the help of larger and larger amounts of artificial sweeteners. Artificial sweeteners, however, do increase your desire for sweet things, and for most people, seem to eventually increase rather than decrease sugar intake.

Artificially sweetened products along with a strict diet may allow some short-term weight loss, but the bottom line is this: **Artificially sweetened products lead to long-term weight gain, whether or not calories are reduced.** In fact, the reduction of calories serves as a major stimulus to make your body metabolically more efficient, to gain weight more quickly, and to ultimately store more fat. The sweet taste may also contribute to weight gain, and just as with sugar, can do it with no increase in caloric intake.

Besides increasing your desire for sweet-tasting foods, artificial sweeteners seem to interfere with your enjoyment of good, wholesome foods that are not as sweet. As long as you continue eating highly sweetened food or drink on a regular basis, you will want and enjoy foods with high-sugar (and often high-fat) content. You will then not enjoy the taste of the unrefined wholesome foods ideal for good health, and they will not taste sweet enough to satisfy you. Very few people will continue on a long-term basis to eat food that is not satisfying and that doesn't taste good to them. The use of artificial sweeteners thus seem to interfere with good food choices.

In addition to promoting weight gain and contributing to various degenerative diseases, artificial sweeteners contribute to many other problems, as discussed in Chapter 5. Many of these symptoms are caused by an imbalance in brain chemistry. The new sweetener aspartame seems particularly prone to cause brain chemistry disturbances due to the amino acids released when it is metabolized. Some of these amino acids are either converted into brain chemicals, or compete with other amino acids that would otherwise form brain chemicals.

We are deeply concerned by the number of people we are seeing in our office who are suffering from depression, sleep disturbances, irritability, anger, trouble thinking clearly, trouble concentrating, trouble coping, and other problems that seem to be the result of altered brain chemistry. Whether these problems are becoming more common is hard to determine, but they seem to be. We are seeing more of these problems in our medical practice than we used to. The suicide rate among teenagers has soared to alarming levels in recent years. Millions of adults, as well as young people, are turning to drugs in an attempt to feel good. Chains of psychiatric hospitals are springing up all over the country. There is a tremendous amount of human suffering and loss of productivity resulting from these emotional problems.

It is now quite clear that the balance of these brain chemicals (neurohormones) is dependent on the nutrients in the diet. If your diet lacks the right balance of amino acids and other nutrients that influence neurohormone production, you will not end up with the optimal balance of brain chemicals. One of the very important and yet largely overlooked factors in the balance of brain chemistry is the role of eating to satiety. **When you eat enough food to completely satisfy your biological needs and your hunger, you will develop a feeling of complete satisfaction, contentment, peace, happiness, pleasure, and a sense of well-being that you can obtain in no other way.** Eating to satiety on a regular basis seems to be a very important part of obtaining and maintaining an optimal balance of brain chemicals.

It is very obvious to us that people with emotional symptoms can often be helped simply with dietary changes. Over the last five or more years since we have been teaching the eating and weight-control principles discussed in this book, we have seen hundreds of people overcome these problems. Many have been able to stop taking tranquilizers, sleeping pills, and antidepressant medication. The improved quality of life has often been dramatic.

The high quality of life produced by a good diet is in marked contrast to that of many artificial sweetener users. Of particular concern are the young diabetics whose parents, acting out of love and concern, allow them to have artificial sweeteners. They reason that "life is tough enough for my diabetic child without being deprived of the pleasure that comes from sweet-tasting foods and drinks." These children then develop a continual yearning and longing for foods they shouldn't have, and they don't find the foods which they should eat very enjoyable or satisfying.

It is tragic how many people—diabetics and non-diabetics alike—completely sacrifice the satisfaction and contentment they can get from healthy foods in exchange for a few fleeting moments of pleasure from highly sweetened products. Many people suffer from hunger and sweet cravings almost every waking minute, except for a few moments of partial relief they get from a diet drink. The major reason we wrote this book was to explain that *there is a better way.*

If you've been trying to eat a healthy diet, but have hung onto artificial sweetners, get rid of them. Eat healthy foods—and eat enough of them to completely satisfy your needs. Before you know it, you'll be enjoying them—you'll find them completely satisfying, and you will experience a sense of fulfillment, peace, and contentment that you might not have known was possible.

Section 2 of this book tells you how to successfully control your weight and your eating for optimal health. The last few pages at the end of the book provide information about products and services that might also be helpful.

1. Cranmer MF. Saccharin: A report. Ed: Scherr GH. American Drug Research Institute, Inc. 1980; pp. 14.

2. Rhein RW, Marion L. The Saccharin Controversy: A guide for consumers. New York, NY: Monarch Press 1977; pp. 98. Statement by the Calorie Control Council attributed to Dr. Steven Scheidt, undated.

SECTION II

CHAPTER SEVEN

THE CAFFEINE AND CHOCOLATE ISSUE

If you've been swilling lots of diet drinks, chances are that you are also getting plenty of caffeine on top of the artificial sweeteners.

What's so bad about that?

In some cases, the caffeine might cause unpleasant symptoms. This may especially be a problem if you are using artificial sweeteners at abuse levels.

Do you have any idea how much caffeine you consume in an average day? Do you know what the effects of caffeine are? The amount of caffeine may vary a great deal, depending on the source of the raw materials and the processing techniques, but here are some averages:[1,2,3,4]

PRODUCT	AMOUNT	CAFFEINE CONTENT
Coffee - instant	1 cup	106 mg.
Coffee - percolated	1 cup	118 mg.
Coffee - drip	1 cup	179 mg.
Coffee - decaffeinated	1 cup	2 mg.
Tea	1 cup	50 mg.
Cola drink - regular	12 oz.	60 mg.
Cola drink - diet	12 oz.	60 mg.
Chocolate	1 oz.	20 mg.
Chocolate milk (cocoa)	1 cup	up to 40 mg.
Anacin®, Bromo Seltzer®	1 pill	32 mg.
Excedrin®	1 pill	60 mg.
NoDoz®	1 pill	100 mg.

Drug-like effects of caffeine.

Caffeine is definitely a drug, and it has a number of effects on those using it. Most people are aware that caffeine interferes with sleep. It is usually considered to be a central nervous system stimulant, but may have variable effects on behavior and performance. In some cases, it may improve energy, shorten reaction time, and enhance athletic performance.[5,6] In other cases, it may cause hand tremor, impair fine motor coordination,[7] and cause jitters and nervousness.[8] It may improve intellectual function in some people, but impair it in others.[9,10]

Caffeine has been shown to cause increased secretion of gastric (stomach) acids,[11] which may contribute to gastric ulcers. Other substances found within caffeine may stimulate acid secretions even more than caffeine.[12] Decaffeinated coffee may thus also aggravate ulcer symptoms.

Caffeine has a diuretic effect, causing an increased output of urine. Caffeine stimulates heart muscles, can cause the heart to beat faster, and can increase the volume of blood pumped by the heart. Caffeine has many effects on various structures throughout the body, most of them being rather minor.[13]

Addiction problem.

One of the most well-known problems associated with caffeine is its potential for becoming addictive. Most of those who use it several times daily eventually become addicted to it, and begin to have withdrawal symptoms when they haven't had caffeine for a few hours. These withdrawal symptoms can vary somewhat from person to person, but usually include headaches, irritability, fatigue, shaking, and depression.

When caffeine addicts consume coffee or diet drinks that contain caffeine, it often exerts a calming, relaxing effect on them, since it stops withdrawal symptoms. It may even help them to sleep better, since they might otherwise go through withdrawal during the night and not be able to sleep as a result.

The potential of caffeine to cause irritability and central nervous system stimulation, and the ability to calm these effects in those who are addicted, make it very difficult to study the effects of caffeine. One of the typical features observed in caffeine users is the mood and energy swings from one time of the day to another. You might have plenty of energy and be in a great mood with a general sense of well-being for a few hours after ingesting caffeine. A number of hours later, though, you might feel very tired, depressed, and irritable.

What do you do? Head for more caffeine. The result? Your good feelings return—only to be followed by another drop in energy and a creeping depression.

A good share of our population use caffeine-containing products on such a regular basis that they are addicted. When they try to stop using caffeine for a few days to see if it is causing any of their health problems, the symptoms frequently get worse, and they may conclude that caffeine isn't to blame. The only way to identify symptoms caused by an addictive substance like caffeine is to stop it long enough to get over the withdrawal symptoms, and then evaluate. Symptoms that may have bothered you for years may disappear completely—almost certainly the result of your addiction.

Headaches are common among caffeine users. Headaches can be caused both by the caffeine itself and by the withdrawal from it. Headaches seem to come on very gradually after many months or even years of using caffeine, and seem to get gradually worse. Because of this gradual onset, many caffeine users don't relate the headaches to caffeine. The headaches may start shortly after caffeine ingestion, or they may not start for a number of hours— and sometimes not until the next day. A headache present first thing in the morning that goes away shortly after caffeine ingestion is typical of a withdrawal headache.

Headaches caused by caffeine can follow a number of different patterns. They can be almost anywhere on the head, and may be typical of migraine, tension headache, or may be generalized, with no clear-cut features.

Effect on Weight: Caffeine has been used in a number of over-the-counter diet pills, partly to overcome the chronic fatigue associated with dieting. It has also been shown to increase the metabolic rate,[14] which has been thought by some to help burn off extra calories. It has been shown, however, to interfere with glucose metabolism, raise the blood sugar, and increase fatty acid levels.[15] These effects might be associated with increased insulin resistance, which would be expected to interfere with weight loss. We are not aware of any clear, long-term effect of caffeine on body weight that has been established through scientific studies.

Fibrocystic Breast Disease has been associated with the use of caffeine and other similar chemical compounds called xanthines. These chemicals are also found in chocolate and in certain medications used for asthma. Some women develop painful nodules in the breasts, which are often worse at certain times of the menstrual cycle. These breast lesions have no apparent cancerous potential, but may be confused for cancer and lead to unnecessary breast surgery. A high percentage of these innocent but painful breast problems can be completely eliminated simply by getting the caffeine and other related chemicals out of the diet.

Reproductive Disorders have been demonstrated in animals ingesting caffeine.[16,17,18,19,20] In humans, high levels of caffeine ingestion during pregnancy have been associated with an increased incidence of spontaneous abortions and stillbirths, breech deliveries, and breathing difficulties at birth.[21,22] One researcher feels strongly that pregnant women should be warned not to consume more caffeine than that found in a fraction of a cup of coffee daily.[23]

Heart Disease has been linked to heavy caffeine use.[24,25,26] It took many years to finally establish this link. One of the first clues was a study done in California among Seventh-Day Adventists. Those who drank coffee had a higher rate of heart attacks than those who did not drink coffee. Incidently, both groups lived considerably longer than average because they usually avoid drinking alcohol, smoking, and eating meat.

It has been difficult to establish a relationship between coffee consumption and heart disease, since there are so many other variables involved. For example, it was clearly shown in a study involving more than 1,100 physicians that heavy coffee drinkers also tended to be cigarette smokers. Those drinking 5 or more cups of coffee daily were 2.8 times more likely to have heart problems than those who drank no coffee at all.[27] In studies such as this, some investigators have wondered whether it is the caffeine or the smoking that causes the heart problems.

Some researchers have shown that heavy coffee drinking contributes to heart disease by causing higher levels of total cholesterol and other lipids known to contribute to heart disease.[28,29,30] Most of the studies dealing with caffeine have concentrated on coffee consumption. It is certainly possible that other components of the coffee besides the caffeine might contribute to the heart problems. By the same reasoning, it is also possible that other components of cola drinks may be a problem for some people.

CHOCOLATE

Chocolate, a favorite food for many people, has also been shown to cause addiction—as any "chocoholic" can tell you. But there are reasons why so many people love chocolate. Besides caffeine, chocolate contains a number of other chemicals, including phenylethylamine. This substance is also produced naturally in the human brain, and is believed to be produced in higher amounts when people fall in love.[31] The effects of phenylethylamine may account for the "feeling good" state that so many people seem to experience from eating chocolate.

Besides caffeine and phenylethylamine, chocolate contains a number of other substances that are very appealing. Most chocolate products contain a great deal of sugar, and recently manufactured chocolate products may contain artificial sweeteners—which may them very sweet. Chocolate products also contain fat—and, in some products, as much as 94 percent of the calories are derived from fat.[32]

The taste of chocolate is also very appealing. This combination of high fat, high sugar, and good taste has been utilized by researchers who have given chocolate chip cookies and milk chocolate to rats in order to make them fat for experimental purposes.[33] Needless to say, chocolate also works well for making people fat!

Chocolate perhaps has a greater potential for addiction than any other food (with the possible exception of alcohol). Caffeine addiction-withdrawal symptoms can often be appeased by chocolate, since it contains caffeine, but a chocolate withdrawal reaction can be eased only by eating chocolate itself. Sometimes the addiction to chocolate can be devastating. A young women in our practice developed severe headaches within hours of eating chocolate. The chocolate would cause vomiting and visual disturbances, and she would spend almost two days in bed totally incapacitated each time she ate chocolate. In spite of it all, she couldn't stop herself from eating it at least once or twice a month!

Adverse reactions from chocolate seem to be very common. Babies often develop diarrhea, colic, or skin rash when their nursing mothers use chocolate. Many children can tolerate smal amounts of it, but seem to react at Easter or Christmas when they eat larger amounts of chocolate. A wide variety of symptoms can occur among adults who eat chocolate, but headaches seem to be the most common symptom. Headaches can be caused by chocolate withdrawal, from the acute effects of the chocolate—and, in a surprising number of people, headaches occur from both eating it and withdrawing from it.

So take a good look at yourself. Are you a caffeine or chocolate junkie?

Cut back—or, better yet, cut it out! Not only do you risk unpleasant side effects and the danger of heart disease, but you are jeopardizing your chances for a nutritious, well-balanced diet as well!

CHAPTER EIGHT

NEW WEIGHT LOSS TECHNIQUES

Remember Chapter One, where you learned that being fat didn't necessarily mean you are eating too many calories?

In this chapter, we'll explain some exciting new techniques of weight loss based on these scientific concepts. To begin, let's take a brief look at the principles upon which these new techniques are based.

The amount of fat stored in your body is closely controlled by centers within your brain which we call the **fat thermostat**. The fat thermostat "chooses" how much fat to store, and then attempts to keep your fat stores at this chosen level (setpoint). How? The fat thermostat directs the following:

1. **Hunger** is a sensation you experience when you have a biological need for food. When you have eaten enough food to satisfy your requirements, you then develop a feeling of **satiety** or satisfaction, and you stop eating. In a few hours, you will again begin to experience a sensation of hunger, which gets progressively stronger until you eat again. Various neurohormones within your brain determine your degree of hunger or satiety. To a large extent these neurohormones, controlled by the fat thermostat, control your eating.

2. **Energy-wasting systems** in your body tend to "burn off" any extra calories that are over and above your basic needs; they enable some people to eat huge amounts of food and yet successfully keep their weight down. Other people gain some weight if they eat more than they need, but they will not usually gain very much—and overeating usually stimulates these energy-wasting systems to become more active. They will then have to eat considerably more than their basic requirements to continue gaining, or even to maintain the weight that they gained. If they eat at the level directed by their natural hunger drive, their setpoint will usually cause the weight to return to normal levels.

3. **Energy conservation systems** in your body tend to resist loss of fat stores. If you eat less food than you need for your energy needs, these energy conservation systems are stimulated. What happens then? You become progressively more capable of maintaining your fat stores, even when you eat relatively small amounts of food. As soon as you resume eating a reasonable amount of food, the enzyme systems responsible for storing fat are activated to quickly regain any lost fat.

The fat thermostat seems to be able to control your fat stores and your body weight at the setpoint level almost in spite of how much food you eat. But the setpoint is not permanently fixed at one position. It can change, depending on various factors. The key to effective, permanent weight control is knowing what factors influence the fat thermostat. Stop doing those things that raise the setpoint and start doing those things that will lower the setpoint. Your body weight will then decrease to the new setpoint level, and will maintain itself there with little effort.

The fat thermostat is a starvation defense that helps you survive. Your ancestors would not have survived prolonged periods of starvation without ample stores of fat and the ability to utilize those fat stores efficiently. In response to reduced food intake, whether because of a real famine or from a reduced-calorie diet, the fat thermostat directs metabolic changes that conserve energy and protect the fat stores. As soon as food is again available—or as soon as the dieter gives in and eats in response to the increasingly strong hunger demands that result from dieting—any lost fat is quickly restored. In response to this episode of perceived starvation, the fat thermostat often "chooses" to store even more fat to give better protection from any future period of food shortage. The result? You get fatter even if you eat exactly the same as you did before the diet.

In addition to protecting you from starvation, the fat thermostat also seems concerned about your **mobility.** Your ancestors may have survived only if they could move fast in order to escape from enemies, capture game, or migrate long distances to find food.

These two survival features are not compatible: the more fat you store, the less mobile you are; the more streamlined and mobile your body is, the less fat it can store. The fat thermostat, then, seems to "choose" an amount of fat to store that gives you reasonable protection against starvation and, at the same time, a reasonable degree of mobility. It then modifies the amount of fat stored based on what it perceives your needs to be. If plenty of food is available and you are very active, you store less fat to increase your mobility. If you periodically miss meals or reduce food intake, especially if you are inactive, your fat thermostat will try to store more protective fat.

If this concept of obesity is correct, then you can become excessively fat in only three ways:

1. Your fat thermostat "chooses" to store more fat than is considered ideal by current fashion standards.

2. Something goes wrong with the metabolic processes through which you produce and store fat, burn fat, or waste excess energy.

3. Your food intake regularly exceeds your needs and the ability of your energy-wasting systems to eliminate the extra calories. Until recently, this third mechanism was thought to be the only factor responsible for weight gain. It may actually be relatively unimportant, and may be a factor only when something is wrong with the enzyme systems responsible for energy wasting.

What steps lead to effective weight control, and how does each step help to control weight? And how can you implement each step? The following is a brief summary of the concepts taught in *How To Lower Your Fat Thermostat* (Vitality House International, Inc. 1983). Let's take a look:

1. **Exercise regularly.** Exercise tends to lower the setpoint of the fat thermostat. It also increases the levels of fat-burning enzymes. Burning fat through muscle activity mediated by these enzyme systems is the most effective way of getting rid of your fat.

Exercise also lowers insulin resistance and improves sugar as well as fat metabolism. It increases the metabolic rate (the rate at which your body uses energy) during exercise and for a number of hours afterwards. It maintains your muscle mass, which in turn helps you burn more fat and calories. Exercise also changes various neurochemicals, makes you feel good, and gives you a sense of well-being.

These benefits from exercise occur optimally if you exercise at the right rate. If you don't work hard enough, you don't get all of the benefits. If you work too hard, your muscles tend to use sugar instead of fat for fuel, and you develop the sugar-burning enzyme systems instead of the fat-burning systems. Exercise should get your heart rate to within 70 to 85 percent of your maximum heart rate (which is 220 minus your age). For example, if you are 40 years old, your maximum heart rate would be 220 − 40 = 180. Your ideal training heart rate would be between .70 × 180 = 126 and .85 × 180 = 153 beats per minute. Take your pulse during exercise for 10 seconds, multiple that number by 6 to get the beats per minute. If your pulse is too slow, speed up—if it is too fast, slow down.

You also need to exercise for the right length of time. Although you can get cardiovascular benefits from only twelve minutes of continuous exercise, the changes useful for weight loss don't begin until you have exercised steadily for at least thirty minutes—and you may need to exercise for up to sixty minutes to get the maximum benefit.

You should exercise every day, or for at least five or six days out of each week. Exercising three times a week is of some value, and may help your basic fitness level, but exercising more often is much more helpful for weight loss.

You should do exercises that involve large muscle groups in a steady, rhythmic way. Walking is excellent, and so is moderate jogging, swimming, cycling, aerobic dancing, using a mini-trampoline, or comparable activity.

WARNING: Before you start an exercise program, consult your physician to make sure that your health will allow you to exercise and get any pointers for getting started in your particular situation.

Start your exercise program at a very modest level (perhaps only ten to fifteen minutes of slow walking at first) and gradually build up your time and your pace. If you are not already fit, do not start right out with running. Start with walking, and gradually work into running only after several weeks—or even months—of walking. Don't overdo, since your muscles and joints need to get used to your increased level of activity and "toughen up" to avoid injury. If you go too fast or work too long at the beginning, you will have an unpleasant experience, and be unlikely to establish a regular exercise pattern. You should feel good during exercise. If you feel excessively tired, you need to slow down and increase your activity more gradually.

Most overweight people who try to start an exercise program, hate it, and give up. No wonder—most exercise programs accompany a reduced-calorie diet regime. Exercising and reducing calories just don't mix: if you cut calories, exercise will only make you feel weak, tired, and awful.

Perhaps the most important key to establishing a regular exercise program is to make sure you are eating enough wholesome food on a regular basis to provide the nutrients and energy you need. If you eat right, start slowly, go regularly, and gradually build up, exercise should become a pleasant experience for you and make you feel better—and you should even begin to look forward to it! Even if you have hated exercise in the past, it can become a very pleasant and invigorating part of your new life.

2. **Eat enough food on a regular basis to completely satisfy your hunger.** Eating breakfast is critical, and you should eat at least two other times during the day as well. Snacking is not only permitted, but even encouraged if you become hungry.

You should eat at least one time during the day until you feel completely and totally satisfied. This does not mean stopping once the edge has been taken off of your appetite. It also does not mean stuffing yourself to the point that you become uncomfortable.

The main reason for eating plenty is to provide all the calories and nutrients that your body needs so that you are no longer triggering starvation defenses. When adequate food is coming in on a regular basis, your fat thermostat will then adjust downward in response to the exercise and other features of the program. Adequate eating will also help cause a number of enzyme and hormone systems to begin working in a "thin mode" rather than a "fat mode." In other words, the systems that worked hard to maintain excessive fat stores will begin to work for you in maintaining lower levels of fat. These enzyme and hormone systems may even be developed to the point that you can maintain a lower body weight in spite of occasional indiscretions during holidays or vacations.

Again, let us stress that eating adequate food is extremely important for providing the essential, protective nutrients that help prevent cancer, high blood pressure, heart disease, diabetes, and other degenerative diseases. In addition, adequate nutrients should stop or prevent many of the unpleasant symptoms that are so common as a result of starvation dieting.

3. **Reduce your intake of refined sugar and other highly sweetened substances.** As has been pointed out, refined sugar for various reasons tends to cause weight gain, even when no extra calories are consumed. Highly sweetened food may raise the setpoint directly.

Sweet taste is a trigger for insulin release, and highly sweetened substances cause excess insulin production. Excess insulin seems to be a major factor in producing excessive fat stores, and insulin also interferes with burning those fat stores for fuel.

Sugars and other highly sweetened substances may cause an addiction or addiction-like state in some people, which results in a strong desire for more sweetened substances every few hours. These people then often eat sweetened foods, which are usually high in fat and sugar, to make themselves feel good and to prevent themselves from going through unpleasant withdrawal symptoms. The withdrawal process seems to be associated with the release of various stress hormones, which may interfere with insulin action and contribute to weight gain.

Reducing the sugar intake might do the trick for some people—but anyone with an addiction or addiction-like problem with sugars or sweets should cut them out completely, at least for several months.

4. Decrease your consumption of fat. High-fat diets have been shown to cause weight gain even though no extra calories are consumed. This might be a survival adaptation that gave our ancestors a better chance of surviving winter. Many animals increase their fat stores in the late fall. When our ancestors ate these animals, the extra fat may have been a signal to store more fat in preparation for winter, when food would be less available. Nuts and some seeds that mature in autumn are also high in fat, which could have also helped in this fat-gaining survival mechanism. In our modern society, most of the flesh-food we eat is high in fat. Even low-fat flesh food like fish and chicken are often cooked in a way that adds extra fat. Since most of us have the same survival mechanisms as our ancestors, this may trigger extra fat storage on a year-round basis.

High fats may also interfere with insulin metabolism, which in turn could lead to more fat storage.

It appears that most kinds of fats in excess will lead to weight gain. Some types of fats may be worse than others for causing weight gain, since fats contain different kinds of fatty acids—and the type of fatty acid determines the type of prostaglandins your body will produce. Some prostaglandins may play a major role in stimulating energy-wasting cycles. In some people, an imbalance of these prostaglandins might be the factor that contributes to excess weight gain. In the future, we hope the role of these prostaglandins in weight control will be better understood, making it possible to include a lot of tasty fat in the diet. At the present time, however, it is best to limit all fats—although some are more healthy than others—as discussed in the last chapter.

Both fats and sugars contribute to excess weight. They are very low in nutrients: sugars contain virtually no vitamins, minerals, or fiber, just empty calories. Animal fats may contain a few fat-soluble vitamins, but refined vegetable fats contain no vitamins or any other kinds of nutrient (except fatty acids)— only calories. If your diet includes much fat or sugar, your control centers may encourage you to eat a large amount of food in order to obtain the essential nutrients. This would boost the number of calories you eat. Those without adequate energy-wasting systems may thus put on excess fat by repeatedly eating high-fat or high-sugar food.

These high-density foods pose another problem. In order to be satisfied, a certain amount of chewing and filling of the stomach is required. Highly refined foods often require little chewing, and slip down into the stomach with little effort. By the time you eat enough of these foods to satisfy your need for chewing, filling the stomach, and producing satiety, you may end up with more calories than your energy-wasting systems can eliminate, leading to weight gain.

Foods that are both high in fats and highly sweetened are very palatable and appealing. This combination has been shown in a number of experiments to be the most effective for causing weight gain. These foods may raise the setpoint, causing your body to regulate your fat stores at a higher level. They may also stimulate high insulin production.

Eating a typical high-fat, high-sugar diet causes perhaps the majority of our population to gain extra fat. We would like to make it clear that the weight gain is not caused by the extra calories in high-fat, high-sugar foods. On the average, overweight people eat no more than thin people, and most comparative studies have actually shown that overweight people eat less. It seems to be the dieting after an initial small weight gain that causes most of the metabolic changes associated with obesity.

Those who begin to cut their caloric intake but who eat the same types of foods start to develop deficiencies in various minerals and other nutrients. This contributes to various metabolic problems, including increased insulin resistance, higher insulin production, and more fat gain. Eating less food may trigger starvation defenses, raise the setpoint, and cause even more fat gain. The many symptoms and serious health problems associated with obesity begin, and become worse as these dieters progressively impair their nutritional status by vacillating between reducing food intake and indulging in high-energy, low-nutrient foods.

5. **Increase your intake of complex-carbohydrate foods.** Vegetables, whole grains, and fruit are high in complex carbohydrates, fiber, and nutrients, and are generally low in fat content and caloric density. These foods require chewing and they fill the stomach—which contribute to satiety. Eating lots of these foods seems to lower the setpoint and cause regulation of body fat at a lower level.

The ideal balance in your diet should be approximately as follows:

Fat .... 10 - 20% of total calories
Protein .... 10 - 15% of total calories
Complex carbohydrate .... 65 - 80% of total calories

Suggestions for lowering fats and sugars and increasing complex-carbohydrate foods will be made in the next chapter.

6. **Increase water consumption.** Water is essential for various metabolic processes to occur and for eliminating wastes. As we have looked at thousands of patient food records over the years, we have noticed that many—if not most—overweight dieters do not drink very much water. This may contribute to some of the health problems associated with being overweight, and may interfere with effective weight loss. You should drink six to eight glasses of water a day—but never more than twenty glasses in a single day. (Drinking too much water can cause problems, too!)

7. **Eat and drink in harmony with your hunger and thirst drives.** If you are hungry, you have a biological need for food. You should eat enough food to completely satisfy your hunger, and then stop eating. If you are thirsty, you have a biological need for water, and you should drink water. Drinking fluids other than water may cause problems for the overweight person. Drinks containing artificial sweeteners or refined sugars contribute to weight gain in ways that have been extensively discussed. Alcoholic beverages are high in calories, low in some nutrients, and tend to be used in an addictive way. Alcohol is metabolized like a fat, and the tendency for it to promote weight gain is well known. Coffee and tea contain few calories, but can clearly be a problem if sugar, artificial sweeteners or high-fat cream are added. The caffeine in these products may also play some role in interfering with insulin metabolism and contribute to weight gain.

Even drinks that have been thought to be very healthy may present problems for the overweight person. Most milk (except skim) contains significant amounts of fat. One or two glasses of low-fat milk would probably be reasonable, but drinking milk with each meal or drinking it when you are thirsty can be a problem.

Juice contains easily digested sugars, has been stripped of the fiber and other nutrients contained in the whole fruit from which it is processed, and provides no chewing. It would certainly be better for the overweight person to eat the whole fruit and obtain the benefits of extra fiber, nutrients, and the chewing and eating enjoyment it would provide.

8. **Decrease your stress.** Stress seems to be a major factor for both causing weight gain and preventing effective weight loss. Your body only has a limited number of ways it can respond to stress. The **fight or flight** response to stress is taught in all basic psychology classes. The stress hormones involved may interfere with insulin and cause fat gain. The setpoint of the fat thermostat may also increase in response to stress. Although this would be a useful response to the stress of starvation, it would be a very undesirable response to other forms of stress. We have seen enough people who gained weight rapidly in stressful conditions—even though they ate less than usual—to believe that this may be a major problem. We have also seen a lot of people who could not lose weight effectively during stressful situations, but who could lose again once their stress levels were reduced.

Excessive stress may also interfere with your ability to live a healthy lifestyle. If you are too involved with problems, you may not have the time or energy necessary to shop wisely, prepare suitable foods, or exercise effectively. Other things in your life may take priority over making the lifestyle changes needed to be thin and healthy.

It is beyond the scope of this book to discuss stress and its management in any detail. Although counseling with a skilled professional may be needed in some cases, there are a number of practical, common-sense changes you can easily make. If you have stress from money worries, it is very easy to cut back on your expenditures, make your old car last a few more years, dress more simply, stay in living accommodations you can afford, and in other ways live within your means. If you are stressed with too many things to do and not enough time to accomplish them, you may need to cut back on your commitments, expect less of yourself, learn to say no, learn to be more efficient, and learn to budget your time more effectively. Various relaxation techniques may be helpful and are easily learned from various self-help books. Relaxing hobbies may be useful.

What to expect with lowering your fat thermostat system.

Be patient, and don't expect rapid weight loss. Many weight-loss programs promise very rapid weight loss. Weight-loss claims of ten pounds in one week, thirty pounds in one month, two pounds in one day, or even claims of weight loss within a few hours are commonly made. Rapid weight loss is mostly muscle and fluid, and only small amounts of fat—if any—are lost.

Weight loss with the system we have described may begin rather slowly. If you have lost muscle from previous episodes of starvation dieting, you may actually gain weight at first as you build back muscle mass. This is healthy and will help you in the long term, because muscle tissue is important for strength, endurance, energy, and fat-burning. Even if you are effectively losing fat, muscle gain may offset any weight loss, and you might even gain some weight at first.

If you have been on many diets in the past and have been living in a state of semi-starvation, it may take some time before the setpoint moves down—it won't readily want to surrender some of your protective fat stores. It also may take a reasonable amount of exercise in order to lower the fat thermostat. It may take some time before you can achieve the fitness level necessary to stimulate your body to make the needed physiological changes for effective weight control.

Although some people take several weeks or even a few months before effective weight loss begins, many others lose weight right from the beginning. Once weight loss begins, it often accelerates as your fitness level goes up and you begin to make metabolic changes that enhance fat loss. (Remember that with reduced-calorie diets, there is rapid initial weight loss that then slows down and eventually stops as soon as your body adapts to starvation.) With this system, women will lose on the average of 1 to 1 1/2 pounds of fat per week, and men will lose on the average of 1 1/2 to 2 1/2 pounds per week.

If you have always gauged your success by the amount of weight you lose, you need to change your thinking in order to stay with the program. You should look at other indicators of success instead of the weight loss. As you lose fat, you should lose inches, and your clothes should fit more loosely—sometimes before you lose any weight. Your energy level should improve, and your ability to exercise should get better. You should start feeling better, and often a wide variety of unpleasant symptoms— like headaches, anxiety, sleep disturbance, depression, arthritis, and digestive disturbances—will gradually disappear. Excessive hunger should diminish, and food cravings should be greatly reduced. You should feel more in control of your eating.

Advantages of the lowering your fat thermostat system.

LOWERING FAT THERMOSTAT	CONVENTIONAL LOW-CALORIE DIETING
Setpoint goes down.	Setpoint: same level or goes up.
Weight loss is maintained with little effort.	Weight tends to be regained: may get fatter than before diet.
Can eat to satiety, only natural between-meal hunger.	Must always stay hungry to keep weight down.
Can eat plenty of nutritious food and still keep weight off.	Must keep food intake low to keep weight down.
Low food awareness, little tendency to binge.	Preoccupied with thoughts of food, tendency to binge.
Various enzyme and hormonal systems tend to normalize and become more like that of a thin person with high metabolic state	Various enzyme and hormonal systems programmed by dieting to conserve fat even better, maintains low metabolic state
Adequate diet protects from developing serious disease.	Reduced food intake, inadequate intake of protective nutrients.
Maintain muscle mass, improved strength, energy, endurance.	Tend to lose muscle mass, get weak, tired, poor endurance.
Free from symptoms associated with starvation effect.	Develop many unpleasant symptoms as a direct result of prolonged inadequate food intake, ie: depression, anxiety, sleep disturbance, irritability, headaches, edema muscle pains, etc.

Factors that interfere with effective weight loss.

Most of the people who follow the guidelines suggested here will eventually lose weight. In spite of their best efforts, however, not everyone will lose as much weight as they want, and a few people will not lose weight at all. We have now identified a few of these roadblocks to successful weight loss, and have been able to help people with these problems. The following factors seem to interfere with effective weight control:

Food Addiction seems to be common in those who have gained excessive amounts of weight. It has been our clinical experience that the repeated cycle of eating an addictive substance followed by withdrawal effects from that substance interferes with effective weight loss. In particular, those with addiction to sugar or chocolate have problems because of the effect that repeated intake of these substances has on the centers that regulate weight. Even foods with few or no calories (like artificial sweeteners and caffeinated products like plain coffee or tea) commonly used by dieters may interfere if you develop an addiction to them.

The mechanism by which addiction interferes with weight has already been described. Help in eliminating a food addiction problem will be given in Chapter Ten.

Artificial Sweeteners are a problem—as explained in detail in the first two chapters, they are responsible for causing weight gain. It has been our clinical experience that they also interfere with effective weight loss. We have seen very few (if any) people who were able to consistently choose and be completely satisfied with good, healthy, low-fat, low-sugar foods while using artificial sweeteners on a regular basis. Because of this observation, we are convinced that they interfere with the ability to follow the outlined program.

There are also a number of other mechanisms outlined earlier in the book that seem to contribute to the weight problems associated with artificial sweetener use. Regardless of the mechanisms by which they interfere, it is very important to stop using them so that healthy, nutritious foods begin to taste sweet enough to be enjoyable and satisfying, to trigger the necessary insulin response, and to signal satiety when you have eaten enough food.

It may take ten days to two weeks after you stop eating highly sweetened foods and drinks before your taste will adjust, but if you are patient, these changes will take place. The nutritious food will not only taste remarkably sweet and enjoyable, but you may even find you prefer it over the problem foods.

Food Allergies may also contribute to weight gain and interfere with effective weight loss. The term allergy in this description refers to the broader definition of allergy as a "condition of unusual sensitivity to a substance or substances which, in like amounts, do not affect others." On a number of occasions we have seen effective weight loss accompany a favorable response to allergy treatment—even when previous determined weight-loss efforts had failed. Very little has been written in the scientific literature about the relationship between allergy and weight problems, but several books written for the lay public dealing with allergy discuss this relationship.[1,2]

Various effects of allergy can affect weight. Addiction or allergy-addiction to foods has been described, and can be a major factor in causing a weight problem. Allergic responses to foods can trigger excessive hunger or cravings, especially for the food involved. Even allergic responses to chemicals such as exhaust fumes, natural gas fumes, and perfumes may trigger excessive hunger. Food allergies may interfere with the satiety signals from the brain, leading to excessive food intake, which may cause weight gain in those with inadequate energy-wasting systems. Allergic reactions may also cause release of stress hormones, which in turn may interfere with insulin and contribute to weight gain. Fluid retention may also result from food allergy.

Identifying and overcoming food allergy may be an important part of overcoming a resistant weight problem for some people. Dr. Alan Levin provides useful information about this problem in his book, *The Type 1/Type 2 Allergy Relief Program.*[3]

Yeast (Candida) Overgrowth may also contribute to weight gain. As described in Chapter Six, yeast normally live in the intestinal tract, and can overgrow when the immune system is altered or when antibiotics kill too many competing bacteria. Many overweight patients suddenly gain weight after an illness or surgical procedure in which strong antibiotics were used. In these patients, there were usually a number of other problems suggestive of yeast overgrowth. Many of the more than 1,500 patients we have treated for yeast overgrowth have been overweight, and we have often seen dramatic improvements in the weight with this treatment.

The exact mechanism by which overgrowth contributes to weight is not clear, but we would like to suggest a number of theories. Yeast produce various toxic substances as by-products of their metabolic processes. One of these products, acetaldehyde, is a potent inhibitor of various enzyme systems. It seems to interfere with enzymes involved in sugar metabolism, and results in excessive hunger and sugar cravings. It may also interfere with sugar storage and the ability to use these sugar stores once the energy from the previous meal is gone. This contributes to weakness, fatigue, and other unpleasant symptoms a few hours after eating, and may be a major factor in hypoglycemia (low blood sugar). Acetaldehyde may also impair fat metabolism by interfering with the utilization of fat for fuel, and by enhancing fat storage. It may also interfere with the enzymes involved in the energy-wasting systems, and may even interfere with the action of thyroid. Many people with apparent yeast overgrowth are clinically like someone low in thyroid: they have low body temperatures, are cold, feel tired, have dry hair and dry skin, are constipated, and gain weight.

We suspect yeast overgrowth in overweight patients who have taken a number of antibiotics, hormones, or cortisone (all of which can contribute to yeast overgrowth) and have many of the following symptoms:

Intestinal gas, bloating, indigestion, and heartburn.
Constipation, diarrhea, or alternating between these two.
Trouble remembering, thinking clearly, trouble
     concentrating, and trouble focusing.
Anxiety, fears, panic attacks, and easily angered.
Depression, fatigue, loss of interest or pleasure.
Muscle stiffness, cramps, and soreness, and pains.
Sugar or carbohydrate cravings.
Feel sick if don't eat regularly.

Further information about identifying yeast overgrowth problems and managing them is found in our book, **Back To Health: A Comprehensive Medical and Nutritional Yeast Control Program** (Vitality House International, 1986).

Premenstrual Syndrome (PMS) is a common condition in women in which various unpleasant symptoms arise in the few days or weeks prior to the onset of menstruation. Sugar cravings, excessive hunger, and weight gain commonly accompany PMS. Some women gain several pounds with each cycle, and won't lose it all before the gain caused by the next cycle. The weight can continue to climb.

PMS appears to be a rather complex disorder or group of disorders. Part of the problem may be hormonal imbalance, since adding progesterone frequently helps. Nutritional deficiencies seem to play a role, since various vitamins and minerals have been shown to help. Yeast overgrowth may be a factor, since treatment for yeast often reduces PMS problems. Allergies may even play a role, since allergy treatment has been helpful.

Psychological Blockers may interfere with effective weight loss. Excess fat storage may serve some subconscious purpose. Even though you desperately want to lose the weight on a conscious level, your subconscious mind may be threatened by weight loss.

For some women who have been molested or had other unpleasant sexual experiences, extra fat may serve as an insulation to reduce the risks of this happening in the future. As weight is lost, and a more appealing figure is achieved, any sexual attention from men may be perceived as threatening by the subconscious part of these women. They may then have a strong urge to binge, or to somehow sabotage their own weight-reduction efforts.

Some men developed a concept of manhood when they were children. A real man may have been perceived as being big, strong, powerful, and in control. When they reached a high weight, they may have subconsciously felt secure and confident— but, on a conscious level, they may not like the way they look with all their excess fat, or they may feel discriminated against because of their weight. As they lose weight, these man may feel at a subconscious level like they are losing their manhood, their power, and their authority. They may find themselves in a dilemma: part of them really wants to succeed, but another part of them is very threatened and attempts to sabotage any weight-loss efforts.

There is clearly a difference between these subconscious psychological blockers—which are only occasionally seen—and the normal response to perceived starvation that occurs when you aren't eating enough. This is associated with a strong desire to eat, and a preoccupation with the thought of food. Almost everyone will eventually give in to these strong drives. This may happen more under stressful situations, but has nothing to do with having psychological problems.

It sometimes takes a professional counselor to help identify and correct existing psychological blockers when they are present. Some people, however, are able to recognize these problems once they are aware that they exist, and may even be able to overcome them on their own and achieve successful weight loss.

In conclusion, there seem to be many different causes of obesity, and many things can go wrong that ultimately lead to excessive weight gain. Although medical science is just beginning to understand some of these mechanisms, enough is known so that most people can now successfully implement these new concepts for effective weight loss. A great deal of research is still needed to identify all of the various metabolic problems which need to be corrected. As more and more of these problems are identified and ways are found to correct them, weight loss methods will become more effective. We hope that eventually everyone with a weight problem can lose it effectively and comfortably and keep it off permanently!

CHAPTER NINE

HOW TO EAT A HEALTHY DIET

How can you start making changes toward a good nutritional program? How can you find the time for food preparation?

If those sound like ridiculous questions, think about today's lifestyles. You can get food—quickly—without even getting out of your car. You can eat three meals a day without dirtying a pan to cook them in. If you choose a certain category of food, you don't even have to dirty a plate! Frozen foods, microwave foods, canned foods, and easily prepared items are more the rule than the exception. Many people limit their cooking to items that come out of a box, bag, or can.

We have an overabundance of food choices, yet as a society we are undernourished. We have drifted away from grains, vegetables, legumes, and fruits. We eat too much meat and processed meat products, and we drown everything in salt and sugar. The result? Obesity, high blood pressure, heart disease, diabetes, and cancer.

The average American diet contains entirely too much fat (estimated at 40 to 50 percent); too much sugar (125 pounds per person per year in the United States), and other refined carbohydrates; too much sodium (salt); and generally too much protein. As we have analyzed numerous patient diet records, it is not uncommon to find that a patient ate only one or two vegetables during a four-day period. We find that, in all too many cases, a typical breakfast consists of either no food at all, coffee, or a 32-ounce caffeinated soda pop or diet soda pop. Teenagers often stop at the local convenience store on the way to school for a large soda pop and a doughnut or a fried fruit pie. If a 12-ounce can of soda contains ten to twelve teaspoons of sugar, figure out how many teaspoons of sugar are in a 32-ounce or 44-ounce soda. Those who eat at home rarely fare much better—they often eat a bowl of a sweetened cereal loaded with up to 15 grams of sugar per ounce of cereal. Think about that: up to half of the total calories coming from refined sugar!

Thanks to research, we know that diet—the way we eat—has a tremendous impact on health. It has been estimated that 50 to 60 percent of all cancer is directly related to the diet. One out of ten women will get breast cancer during her lifetime. In Japan, the rate of breast cancer is only one-fifth the rate of that in the United States. But if Japanese women migrate to the United States, their risk factor then increases to one in ten. Why? They change their diet. Colon cancer is another cancer that is directly related to the lack of fiber and bulk eaten in the typical American diet.

Coronary heart disease is very common. According to the American Heart Association, more than 600,000 Americans die each year from heart disease, and an additional one million suffer a nonfatal heart attack. High blood cholesterol levels are a risk factor in heart disease. It is estimated that fewer than one-half of all American adults have a healthy cholesterol level. Blood cholesterol levels can be reduced with low-fat diets. High-fat diets have also been linked to the development of autoimmune diseases such as systemic lupus erythematosus and diabetes. High blood pressure has been lowered merely through a diet of less sodium and fewer fats, particularly fewer saturated fats.

For years, foods were refined, and the vitamins and minerals that were removed were calculated and then added back. Those that were associated with the health food industry were busy admonishing people to eat whole-grain products. Many nutritionists felt that it didn't really matter whether the product was whole wheat or white, since scientists had added back all of the lost nutrients. In recent years, however, we've realized the importance of fiber. Groups like the American Heart Association, American Diabetes Association, and the American Cancer

Association are all prescribing the same type of diet: a diet low in fat and high in complex carbohydrate.

Let's take a closer look at recommended diet components.

FATS

Some fats are needed in the diet; without any fat at all, the diet would be so terribly unpalatable that we probably wouldn't want to make the necessary changes to achieve it. The goal is to eat between 10 and 20 percent of your calories in fat. This amount gives you a buffer zone so that you can still eat small amounts of meat, fish, poultry, eggs, or dairy products. One of the easiest ways to accomplish this goal is to eat less animal protein and more plant protein, such as dried beans, potatoes, brown rice, vegetables, and pasta.

Fats provide more than double the number of calories as provided by the same amount of protein or carbohydrate. There are nine calories in a gram of fat, compared to four calories in a gram of protein or carbohydrate. Fats have been implicated in obesity. In one study, two groups of rats were fed exactly the same amount of calories; one group received their calories from a typical American diet of 40 to 50 percent fats, and the other group received their calories in the form of low-fat rat chow. Exercise for both groups was the same. The group on the high-fat diet became significantly fatter than the other group, even though the total calories were the same.

Fats add an appealing taste to foods, but too many are used in the average diet. Fats are blended to improve the taste of foods. In fast food restaurants, it is not uncommon for a special blend of fats to be used in deep-fat frying. Unfortunately, the oils that seem to be the most popular are not necessarily the best from a health perspective. Restaurants and food manufacturers must sell a product to stay in business, so they prepare foods to appeal to the taste and demand of the population. Their sole purpose is to sell, not to cater to the health of a nation. If we, as smart consumers, stopped buying their products, their direction would change and products would become healthier. We could begin by not purchasing anything containing coconut or palm oil.

To understand why, you need to understand a little about the various kinds of fats. Saturated fats are fats that are hard at room temperature. These include animal fats, such as lard, chicken fat, and butter. Cheese and dairy products are high in saturated fats. But two oils—coconut oil and palm oil—are also both saturated fats. They are found primarily in processed foods because they are inexpensive.

Unsaturated fats are oils that are not hard at room temperature (with the exceptions of coconut oil and palm oil). Even in this category, there are several delineations. Mono-unsaturates include oils like olive oil and peanut oil. Polyunsaturated fats include safflower, sunflower, corn, and soybean oils. The polyunsaturated fats also consist of two categories - those that are partially hydrogenated and those that are cold-pressed. Just about every oil on the supermarket shelf is partially hydrogenated, unless the label specifically states cold-pressed.

Why does all this matter? Because the kind of oil you use is important!

At the beginning of the century, oils were extracted by a large roller that pressed the oil out of the seeds. People soon learned, however, that the oils went rancid quickly, giving them a short shelf life without refrigeration. Hydrogenation caused the natural "cis" form of the fatty acids to change into a "trans" form, which is harmful to the body. But essential fatty acids found in cold-pressed oils—which include soybean, safflower, linseed, sunflower, and walnut oils, promote the formation of healthy prostaglandins, types of hormones necessary to the immune system.

Prostaglandins are produced from fatty acids by every cell in the body. Prostaglandin hormones are derived from fatty acids. Warm weather plants and animals have higher levels of Omega-6 fatty acids, whereas cold weather plants and animals have higher levels of Omega-3 fatty acids. Cold-pressed oils contain Omega-6 and/or Omega-3 fatty acids, which are essential to health but deficient in most American diets.

Prostaglandins are critical for health, because they are involved with reproduction, fertility, inflammation, immunity, and intracellular communication. Any of these processes can break down without enough essential fatty acids. In addition, some of the chemical conversion steps necessary for altering prostaglandin levels are interfered with by the presence of saturated fats, alcohol, and various vitamin and mineral deficiencies.

"Bad" prostaglandins result from animal fats, hydrogenated fats, and hydrogenated oils (such as margarine, vegetable shortening, and commercial oils). Bad prostaglandins also result when oil is overheated in cooking or deep-frying. They contribute to pain, inflammation, arthritis, and allergy.

Fish oils have recently come into the fat picture. Fish oils are rich in unsaturated fats called Omega-3 fatty acids, and they actually change the chemistry of the blood in a way that lowers the likelihood of heart disease. Two six-ounce fish meals each week is probably the minimum amount needed in the diet. Not all fish contain the same amount of Omega-3. Those high in Omega-3 fatty acids are anchovies, butterfish, herring, lake trout, lake whitefish, mackerel, sablefish, salmon, sardines, sturgeon, and tuna.

Some of the benefits of Omega-3 fatty acids include lowering of cholesterol levels, lowering of levels of triglycerides, altering of the blood chemistry so that blood clots are less likely to form, reducing blood pressure, slowing the process of atherosclerosis (hardening of the arteries), helping prevent inflammatory diseases like arthritis and migraine headaches, and helping to prevent certain cancers when combined with a low-fat, high-fiber diet.

If you're not eating enough fish, you might want to supplement your diet with marine oils or eicosapentaenoic acid (EPA). The beneficial effects of fish oil are increased by reducing the amount of other fats in the diet. Unhealthy fats can block the formation of healthy prostaglandins even when enough good fatty acids are present in the diet. You can't take fish oil and go on eating all the eggs, bacon, meat, and butter that you want and still expect health benefits.

CARBOHYDRATES

You often hear the term *complex carbohydrate.* but what exactly are complex carbohydrates? They are good components derived entirely from plant sources. They are composed of simple sugar molecules which are considered to be complex for two reasons:

1. They are linked together in long chains. Sometimes these chains are up to several thousand sugar molecules long.
2. They are bound with fiber and other components of the plant which slow down the digestion of these carbohydrates.

Complex carbohydrates are in vegetables, fruits, legumes (such as dried beans and peas), lentils, and whole grains (such as brown rice, whole wheat, cornmeal, and so on). Complex carbohydrate foods are fiber rich, and fiber has been shown to protect against heart disease, diabetes, several kinds of cancer.

REFINED SUGAR

From a nutritional standpoint, refined sugar doesn't supply any of the nutrients we need to survive. With the population consuming more than 125 pounds per person per year, that's more than 600 unnecessary nutrient-free calories per day. Sugar gives a quick pick-me-up, but that is followed by a low feeling a little while later. That low feeling is often accompanied by irritability, inability to think clearly, and weakness. And there's more—as you eat more sugar, insulin is required to handle the rise in blood sugar, and this increases the storage of fat.

Another important aspect of sugar consumption is that the more you eat, the more you want. You actually have a difficult time enjoying the taste of good healthy foods because they don't taste sweet like sugar sweetened products. Artificial sweeteners have the same ability to promote the desire for sweetness. As we have talked with many patients who originally didn't want many sweets, after they started using artificial sweeteners they found they wanted more and more sweets.

SALT

Salt is the leading food additive in the United States. Americans typically consume about 2 1/2 teaspoons of salt each day, when one teaspoon daily would be more than enough. That causes significant health problems: salt is a major contributor to the problem of high blood pressure.

The taste for salt is partially an acquired one. You have the ability to enhance your desire for salty foods, and you also have the ability to reverse it and retrain your tastes to enjoy less salty items. Blood pressure will often drop with the elimination of salty foods. Prior to the era of processed foods, diets were originally low in sodium, but had sufficient amounts of potassium (a mineral that protects against the development of high blood pressure), which is ample in fresh fruits and vegetables.

A breast of chicken is low in fat, low in sodium, and high in potassium. There is a good balance of sodium and potassium in chicken. Chicken contains about 20 to 25 percent of its calories in fats, about 250 milligrams of potassium, and less than 100 milligrams of sodium. If you go out and buy this same piece of chicken in a fast food outlet, the sodium content is from 1000 to 1200 milligrams, it contains the same amount of potassium, and its calories from fat double because it is coated and deep-fat fried under pressure, which forces more fat into the chicken.

Now you know some of the changes that need to be made. What is left to eat?

It's time to get back to basics! Current trends in nutrition are unhealthy, and changes in lifestyles must be made. We have reviewed some of the horrors of modern nutrition, and now we want to give you a logical approach to making the necessary changes. You don't have to be overweight, hypertensive, arthritic, diabetic, or have heart problems in order to benefit from the following guidelines. Young and old alike need to eat this way.

When making dietary changes, do it in phases. Why? You will be overwhelmed and discouraged if you try to do everything at once. The final goal is to decrease fats to 10 to 20 percent of the total calories, increase complex carbohydrates, eat a moderate amount of protein (10 to 20 percent of calories), reduce sugars substantially, and reduce salt intake.

What about vitamin supplements? These, too, have found their way into the average American diet. As people begin to eat poorly, they don't feel well, and they look for additional answers to their problems. Vitamins are helpful in some situations, but they do not replace the need for plenty of good, nutritious, unprocessed foods in adequate amounts. The importance of eating an adequate amount of calories cannot be stressed enough. You cannot get the necessary vitamins and minerals if you are restricting your food intake. The goal for women should be at least 1800 calories each day, and men should eat at least 2200 calories each day.

Phase I (Two Weeks)

1. Reduce foods high in cholesterol and saturated fats. Eat red meats no more than two to three times per week, and eat more poultry.
2. Avoid fried foods; try alternative cooking methods, such as baking, broiling, or sauteing in a nonstick skillet.
3. Eliminate all sugars and artificial sweeteners.
4. Eliminate foods that are high in salt (such as potato chips, processed meats, processed foods, and so on).
5. Eat at least two to four cups of vegetables, two to four fruits, and two to four grains (cereal, bread, pasta) each day.
6. Eat fish at least two times per week.
7. Eat no more than two eggs each week.
8. If you are using whole milk, try 2 percent instead. If you are using 2 percent, try 1 percent or skim milk.
9. Use low-fat dairy products only.
10. Use high-fiber breads (100 percent whole wheat).
11. Drink water for your liquid.
12. Start an aerobic exercise program (such as a program of walking) for fifteen to twenty minutes five times per week.

Phase II (Two Weeks)

1. Eat meat or poultry once per day only. Use more fish.
2. Use less spreads (butter, margarine, mayonnaise) and less cheese (except for skim milk cheese and those low in fat, such as St. Otho® Cheese). Limit cheese to two times per week.
3. Leave the salt shaker off the table (no added salt beyond cooking). Try a reduced sodium salt substitute and reduced sodium soy sauce.
4. When baking with eggs, use the white only (throw the yolk away). Use three whites for every two eggs called for.
5. Use "no salt added" canned products or fresh or frozen vegetables.
6. If you make baked goods, reduce the fat by one-third to one-half.
7. Check labels for coconut oil and palm oil, and eliminate these products entirely.

8. Eat two to four cups of legumes (such as dried beans and peas) each week. If you will use beans in small portions but more frequently, you will probably be able to tolerate them with no gastrointestinal problems.
9. Drink more water each day (six to eight glasses daily).
10. Do aerobic exercise thirty minutes per day five times per week.

Phase III

1. Decrease the salt used in cooking.
2. Use skim or 1 percent milk.
3. Don't use the drippings in gravies.
4. Do aerobic exercise forty to fifty minutes per day five times per week.
5. Rather than drinking fruit juices, eat the whole fruit.

POSSIBLE MEAL MODIFICATIONS

In order to accomplish these goals with yourself or your family, the food must taste good. People don't care if it is high-fat, medium-fat, or low-fat if it tastes good! The key is to find new recipes or adapt old ones to fit these guidelines. You must also remember that there may always be a difference between what is preferred and what is healthy. But as changes are made, healthy foods become not only tolerable, but enjoyable. It will be hard for you to be successful if you are constantly longing and yearning for foods that don't fit in the healthy category. Remember—you don't have to eat "weird" to be healthy!

For Breakfast:

1. Use a no-sugar type of whole-grain cereal, hot or cold, using fresh fruit or dried fruit (such as raisins) as a sweetener and served with skim milk or 1 percent milk. Instead of butter and milk on hot whole-grain cereals, use cinnamon and fruit butter.
2. Use whole-grain toast with unsweetened applesauce, fruit purees, or fruit butters instead of butter or jam. Try using one-half the amount of butter you usually use if you have a sugar addiction. If you have no sugar addiction problem, use one-half to one teaspoon of a reduced-sugar jam on your whole-grain toast, and serve it with a whole fresh fruit.

3. Use plain low-fat yogurt or low-fat cottage cheese topped with fresh fruit and a slice of whole grain toast or a whole-grain muffin.
4. Eat bran muffins using whole-grain flour, no added salt, egg whites only, skim milk, and substituting 2 tablespoons of oil for whatever amount is called for in the recipe. These principles apply in any muffin recipe. You can also substitute some, or all, of the oil called for with unsweetened applesauce.
5. Try French toast, pancakes, or waffles using egg whites only, whole grains, skim milk, no added salt; use an unsweetened fruit juice that has been thickened with cornstarch in place of syrup or jam.
6. Bake extra potatoes the night before. Saute onions, green peppers, and the sliced baked potatoes in a nonstick skillet. Serve with sliced fruit.
7. Possibly try an untraditional breakfast, such as some of last night's low-fat, low-salt dinner; a sandwich made of tuna, chicken or turkey; or a bowl of homemade soup.
8. Sometimes a mobile breakfast is called for. Take two homemade bran muffins and a banana, and you can eat on your way to work.
9. Use fruit sauces in place of maple syrup or high sugar jams. Thicken juice-packed crushed pineapple with cornstarch. Mash berries or puree in a blender and add unsweetened pineapple or apple juice to increase the sweetness. Cook fresh apples until tender and add cinnamon and unsweetened apple juice concentrate to sweeten.
10. Oat bran, used regularly, reduces blood levels of cholesterol. Oat bran can be eaten as a hot breakfast cereal, and is easily mixed into muffins.

Lunch or dinner ideas are interchangeable:

1. Find some new recipes that will fit into the guidelines suggested. Some cookbooks that are worth looking at are:
   a. *Recipes to Lower Your Fat Thermostat* by LaRene Gaunt; Vitality House International, Inc., softback, $14.95.
   b. *Jane Brody's Good Food Book* by Jane Brody; W.W. Norton and Company, softback, $12.95.

c. *The New American Diet Book* by Sonja and William Connor; Simon and Schuster, hardback, $18.95.

d. *Deliciously Low* by Harriet Roth; New American Library, hardback, $17.50.

e. *The Natural High Fiber Life Saving Diet* by Genell Subak-Sharpe; Grosset and Dunlap, paperback, $1.95.

f. *The Pritikin Program for Diet and Exercise* by Nathan Pritikin; Bantam, paperback, $3.95.

g. *Craig Claiborne's Gourmet Diet* by Craig Claiborne; Ballantine publication, paperback, $3.95.

2. Cut all visible fat before you cook your meat.

3. Cook your broths ahead of time and let them cool. You can then skim off all visible fat and proceed with your recipe. This can be done with soups, stews, and so on. You can also put ice cubes in the broth and the fats will rise to the top and harden.

4. On sandwiches, try a little mustard or catsup rather than mayonnaise or butter. Lettuce, tomatoes, sprouts or thinly sliced cucumbers also improve the moistness of the sandwich.

5. In soups or stews, remove and discard all skin from chicken and turkey. If you are roasting or broiling, remove the skin after it is cooked.

6. As you start reducing salt, try other spices and seasonings in your recipes. Fresh garlic and fresh ginger are delicious in stir-frying. Fresh garlic and fresh parsley along with basil and oregano are wonderful in Italian dishes. Try crushing anise in a spaghetti sauce for a sausage-like flavor. Rather than using high-fat sausage, use ground turkey or a small amount of very lean ground beef.

7. Use pureed potatoes, carrots, or beans to thicken cream soups.

8. Substitute plain yogurt for sour cream, and use skim milk for any milk called for in recipes.

9. For salads, add numerous raw vegetables and leave out the cheese and eggs. Try some of the delicious low-calorie dressings available. In homemade dressings, try substituting yogurt and low-fat cottage cheese for mayonnaise, or mix one-half lite mayonnaise and one-half plain low-fat yogurt.

10. It is important not to overcook vegetables and not to cook them in a lot of water. A steaming device is especially helpful in keeping the vitamins and minerals in the vegetables and not leaching them out into the water that is usually thrown away. If possible, use your vegetable water in a soup or stew.
11. Avoid nondairy frozen whipped toppings, as they are high in saturated fats.
12. Look for main dish recipes using dried beans and dried peas, such as pintos and split peas. These are very low in fat and high in complex carbohydrate and protein. You might use them in tostadas, burritos, soups, salads, and even breads and muffins.
13. Use marinara sauce on vegetables rather than a cream or cheese sauce.
14. Marinate potatoes in a no-fat Italian dressing for potato salad.

SHOPPING

1. Look for canned goods that are salt-free, low-sodium, water-packed, juice-packed, or those with no added sugar.
2. When you have a choice, choose fresh fruits and vegetables first, frozen as an alternative, and canned as last choice.
3. Learn to read the labels on the packages. Ingredients are listed according to the amount included. The first ingredient will be the one that is in the greatest quantity, and so on down the list. If sugar is listed as a first, second, or third ingredient, it probably is too much. See if coconut oil or palm oil is listed, and then avoid that product.
4. Many dairy products give you a choice between low-fat or non-fat. If you find a non-fat product, choose that one. Some cheeses, such as mozzarella and string cheese, can be found made from skim milk, while others are made from whole milk. Some registered brand names of low-fat cheese include Countdown®, St. Otho®, Danbo®, feta, neufchatel, and camembert.
5. Don't be deceived by products described as 97 percent fat-free. This figure represents the percentage of fat by weight. What is important is the percentage of the total calories contributed by fat. In order to calculate this, look at the

breakdown of calories, proteins, carbohydrates, and fats. Take the grams of fat listed per serving, and multiply by nine. This will give the total fat calories in that product. Then divide the fat calories by the total calories. This will give you the true percentage of fat. For example, if there are three grams of fat in a serving and you multiply that by nine calories per gram, you get twenty-seven calories of fat per serving. If the total number of calories is eighty-one per serving and you divide that into the twenty-seven calories, you come out with approximately 33 percent fat. So the three grams of fat looks like only a little, but when figuring the percentage of total calories, the fat content is one-third. In 2 percent milk, 37.5 percent of the calories are contributed by fat.

6. Remember that margarine has the same amount of fat in it as does butter. Use these products very sparingly, or eliminate them whenever possible.

EATING OUT

1. If you are asked to bring something to an outing or social event, try taking a low-fat dip with an assortment of fresh vegetables. Even children and young adults enjoy this type of refreshment. Another dish that is enjoyed is fresh fruit. Use pureed fresh fruits as a fruit dip for added interest.
2. For appetizers, choose V-8® or tomato juice, broth soups, fresh vegetables, fresh fruit, and seafood cocktails.
3. For salads, choose fresh fruit and fresh vegetables. Ask for the salad dressing on the side, and add only a small amount. Many establishments now have low-calorie dressings available.
4. For vegetables, choose those that are stewed, boiled, or steamed. Avoid vegetables that are escalloped, creamed, au gratin, fried, or sauteed.
5. For breads, choose whole wheat (when possible), bran muffins, rye bread, corn bread, and a few crackers. Avoid sweet rolls, coffee cakes, danish rolls, croissants, frosted rolls, and biscuits.
6. Limit red meats and avoid bacon; fish, chicken, or turkey are the best choices. Look for roasted, baked, broiled, or boiled. Avoid fried, grilled, sauteed, batter-dipped, breaded, or those with gravy or sauces (tartar sauce). Also avoid goose or duck.

7. It is usually best to choose something other than eggs. One egg is 68 percent fat.
8. Avoid gravy, fried foods, cream sauces, salad oils and dressings, butter, margarine, cream, and bacon.
9. For desserts, choose fresh fruits. Avoid cream or whipped toppings.
10. For beverages, choose water, buttermilk, or skim milk. Avoid cocoa, chocolate milk, shakes, soft drinks, and alcoholic beverages.
11. When choosing restaurant items, avoid menu items described as creamed, cream sauce, buttered, sauteed, fried, escalloped, au gratin, cheese sauce, cheese chowders, breaded, with gravy or sauces, hollandaise, marinated, a la mode, prime, or pot pie.

SNACK SUGGESTIONS

1. Popcorn with Butter Buds® or low-sodium seasoning.
2. Crisped, baked corn tortillas, broken, with salsa.
3. Fresh vegetables served with low-fat dips made with blended cottage cheese.
4. Fresh fruit or water-packed fruit milkshakes made with skim milk in blender.
5. Fresh fruit.
6. Cinnamon-nutmeg applesauce with yogurt topping.
7. Plain yogurt topped with fresh fruit.
8. Non-sugared cereal with skim milk and half a banana.
9. Low-fat, low-sugar muffins.

What it all boils down to is this: there is no replacement for good, healthy foods, and training your tastes to enjoy them is a good place to begin. In the past, people had to make a choice between being thin or experiencing good health. Now you can have the best of both worlds. Low-fat, moderate protein, high-complex carbohydrates, low-salt and low-sugar diet is a healthy diet, and has the potential of eliminating or lowering your chances of developing many catastrophic illnesses!

CHAPTER TEN

OVERCOMING ADDICTION

Addiction to foods—such as caffeine, sugar, artificial sweeteners, and chocolate—can cause a number of problems, including weight gain, headaches, energy and mood swings, and depression.[1] But your addiction doesn't have to be to a "forbidden" food—addiction to reasonably healthy foods can create similar problems!

If you frequently eat in response to an addiction, your eating interferes with good nutrition. We are convinced that getting rid of addictive eating and drinking patterns is an important part of good nutrition and is a vital part of feeling well.

How can you tell if you are battling an addiction? Take a good look at your eating behavior. You are probably suffering from an addiction if you:

- Find it very difficult or impossible to stop consuming a particular food or substance.

- Would do almost anything or pay any price to get that food or substance.

- Feel anxious, irritable, weak, depressed, or have headaches when you have been without that item for a number of hours.

- Feel a lift or boost in energy from eating or drinking the suspected item, and it stops withdrawal symptoms.

To overcome your addiction, try the following steps:

1. **Stop eating or drinking the addictive substance completely.** Sneaking a nibble here and there will only heighten your addiction and prolong the time it takes you to get over it. Read labels, and avoid the problem substance fanatically for several months.

2. **Remove the problem items from your home and from your work place.**

3. **Eat plenty of nutritious food on a regular basis** so that you do not feel hungry and in need of food. This may partially satisfy your addictive drive to eat.

4. **Avoid artificial sweeteners and other highly sweetened foods.** It is important to keep the sweetness level of your foods low so that nutritious foods begin to taste good and satisfying.

5. **Exercise effectively.** Aerobic exercise causes your body to produce endorphins, probably the same substances that are produced by your brain when you eat a food to which you are addicted. Exercise should help ease your withdrawal symptoms and reduce your drive to eat the problem foods. Regular exercise is important, but you can also benefit by doing additional exercise whenever you have a strong desire to eat the addictive food.

   Exercise may also help metabolize some of the stress hormones which can otherwise contribute to unpleasant withdrawal symptoms.

   Exercise may need to be somewhat prolonged (probably in excess of thirty minutes) to be the most effective in conquering addiction. It also needs to be of moderate to high intensity to be effective. Follow the suggestions outlined in the previous chapter—and remember, don't push yourself too hard or overdo. If you are not already exercising routinely, you may wish to establish a regular exercise pattern and become more physically fit before you try to control your addiction. In the meantime, you can gradually reduce the amount of addictive substances that you eat.

6. **Start working on your addiction at a time when you are not under stress.** An accountant would be unwise to try to stop an addiction during tax time. A women should not start just

before she expects a houseful of company for several weeks. Women with PMS should not start at the bad time of the cycle when food cravings are the highest.

7. **Keep busy and keep your mind occupied.** Especially do things you enjoy that are relaxing. Don't produce any extra stress for yourself by trying to do too much.

8. **Avoid places and situations usually associated with your addictive pattern.** If it has been your habit to stop at a bar for a drink when coming home from work, it will be hard to stop drinking if you keep stopping at the bar. If you normally drink a large soda pop while watching a television program, find something else to do instead.

9. **Certain nutritional supplements may be helpful.** The following may ease some of the symptoms and make going through withdrawal more comfortable. Some of these suggested dosages are too high to be safely used for long periods of time. Use these products at these dosages for no longer than one or two months.

Vitamin C	1.000 to 2,000 mg. three times daily
Vitamin E	400 to 800 IU daily
B complex	100 to 200 mg. daily
Vitamin B6*	100 to 200 mg. daily
Evening Primrose Oil	1 to 2 g. three times daily
Magnesium	400 to 600 mg. daily

*Vitamin B6 is in B complex vitamins, but it helps to add more.

CHAPTER ELEVEN
RECIPE SECTION

Although everyone would be better off using very little or no refined sugar, for many people eliminating sugar completely is even more important. A high percentage of the patients we see in our office feel better when they completely eliminate sugar. For some of these people, the difference is dramatic: having even a little sugar after a period of abstinence can make them very ill. Those who are addicted to it find that just one slip causes the craving to return. Diabetics, especially those with insulin-dependent diabetes, should avoid sugar completely.

There's still another group that should avoid sugar: anyone who is trying to control weight. We are not sure exactly why sugar has such a profound effect on weight, but it does. As pointed out in early chapters, sugar can contribute to weight in spite of whether or not extra calories are eaten. We have seen many who, while using sugar, have been unsuccessful in losing weight; once they cut out sugar, they lose five to ten pounds within a few weeks. They often then lose steadily from that point on. Those who have been losing effectively while off sugar often start gaining again once they start eating sugar again.

Switching from sugar to artificial sweeteners doesn't seem to help. It appears to be the excessively sweet taste of sugar and artificial sweeteners that contributes to the weight gain. Like sugar, artificial sweeteners can cause weight gain with no extra calories. Part of the problem with weight gain may relate to the lack of nutrients from eating sugar or artificially sweetened products. Artificial sweeteners may also disrupt the balance of brain neurochemistry, enhance your desire for sweet foods, and interfere with the pleasure and satisfaction you can get from eating nutritious foods.

All of us are in a kind of bind. We are attracted to highly sweetened foods, and yet shouldn't eat them. We are continually exposed to advertisements for problem foods, and to the foods themselves. Those around us are often eating these foods and offering them to us.

What about turning to sugar-free foods? That may not be as easy as it sounds. Why? Many of the recipes for sweet-tasting sugar-free foods have a lot of fat, refined flour, honey or other refined sugar, or artificial sweeteners. Many of the recipes that are nutritionally acceptable just don't taste good, especially if you are used to using sugar or artificial sweeteners on a regular basis. We have searched through many recipe books looking for acceptable alternatives for years, and have found very few.

We believe that the recipes in this section will be very useful for most people, especially for children and those who feel emotionally deprived when others are indulging in "goodies." The recipes in this book really taste good, even to those regularly using sugar and artificial sweeteners. They could be used as desserts and in place of snacks or treats. The cakes could be used for a birthday party, the drinks as an alternative to soda pop, and the popsicle substitutes used to cool off on a hot day. The cookies could be taken to school for a Halloween, Valentine, or end-of-the-year party.

These recipes are all made without refined sugar of any kind, without honey, and with either no fat or low levels of fat. Nutritionally, they are very sound, and should be acceptable for any healthy eating plan. Each recipe is made from nutritious, wholesome ingredients. Please note, however, that each item is not a completely balanced meal. You could not, for example, live entirely on the cookies for long periods of time and expect to get all the nutrients that your body requires.

Keep in mind that despite the fact that these recipes are nutritious and acceptable, they should be used wisely. Problems occur when "goodies" are used as a reward or bribe, when they are offered to children who are hurt or feeling sad, or when they are used to relieve boredom or as entertainment. It is also best not to use the delicious drink recipes on a regular basis to relieve thirst, although they are perfectly acceptable for occasional use. We have previously stressed the importance of learning to eat in response to true hunger, and to satisfy thirst by drinking water, the substance your body is in need of when you feel thirsty.

Although food can present certain problems, food can also play important roles in our lives, in addition to supplying our nutritional requirements. We have stressed the importance of allowing yourself the pleasure of eating to complete satiety on a regular basis to produce the contentment, fulfillment, and sense of well-being that is so important for your individual happiness. Regularly sharing a meal together binds or bonds a family together. Perhaps no other activity gives a family more opportunity to interact with each other and draw closer together. We are concerned by the number of mothers who prepare a meal for their family, and then eat their own diet meal at some other time or location so as not to be tempted by the food the others are eating. With these new recipe items and new weight-control concepts, people who are trying to control their weight, diabetic children, and almost everyone else (except those with allergies to any of the ingredients) will be able to share in eating these foods together.

The following recipes have been adapted to fit the guidelines of no sugar, no honey, no artificial sweeteners, and no salt. The oil content of each recipe has been adjusted as low as possible to still maintain an acceptable product. Many of the recipes call for frozen unsweetened juice concentrate. This means thawed, undiluted juice concentrate. If the recipe calls for apple juice or pineapple juice, then this is in the diluted form.

Following each recipe is nutritional information. **RCU** and **FU** indicate refined carbohydrate units and fat units (designations used for the scoring system in *How to Lower Your Fat Thermostat*). **Cal** represents the number of calories in each serving, and **%Fat** represents the percentage of the total calories in that recipe derived from fat sources. **P, F,** and **C** represent the grams of protein, fat, and carbohydrate respectively for each serving of the recipe. **Na** represents the number of milligrams of sodium in each serving.

Enjoy the items prepared from these recipes, and good foods in general. Look for a forthcoming recipe book from Vitality House International with many more sugar-free recipes. We wish you good eating and good health!

RECIPES

Apple or Grape "Soda"

2 cups chilled unsweetened apple or grape juice
2 teaspoons lemon or lime juice
2 cups chilled club soda

1. Combine juices. Add club soda very slowly. Gently stir. Serve immediately over ice.

Yield: 4 one cup servings

	RCU	FU	Cal	%Fat	P	F	C	Na
Per Serving	0	0	76	0	1	T	19	30

Kiwi Ice

4 kiwi, peeled and cubed
½ cup frozen unsweetened pineapple juice concentrate
1½ cups water
1 tablespoon lemon juice

1. Combine ingredients in blender; blend until smooth.
2. Pour into 8-inch baking pan; freeze until almost firm (about 1 hour).
3. Beat frozen mixture until fluffy. Return to pan and freeze until firm.

Yield: 4 cups or 4 one cup servings

	RCU	FU	Cal	%Fat	P	F	C	Na
Per Serving	0	0	117	4	1	1	32	5

Banana Bites

Bananas
Pineapple juice, unsweetened
Toasted wheat germ or finely chopped nuts

1. Cut bananas into ½ inch bite size pieces.
2. Dip into pineapple juice and roll in toasted wheat germ.
3. Place on a cookie sheet and freeze. Store in plastic bags. Eat these frozen.

Yield: About 5 pieces per banana used.

	RCU	FU	Cal	%Fat	P	F	C	Na
Per Banana	0	0	117	4	1	1	32	5

Baked Apples

4 small baking apples
2 tablespoons wheat germ
2 tablespoons raisins
¼ teaspoon cinnamon
¼ cup chopped walnuts, optional
1 cup unsweetened apple juice

1. Preheat oven to 350 degrees.
2. Core apples, and peel about ¼ of the way down.
3. In baking dish, arrange apples.
4. Combine wheat germ, raisins, cinnamon, and walnuts. Fill centers of apples with mixture.
5. Drizzle apple juice over apples.
6. Bake about 45 minutes or until tender.

Yield: 4 servings

	RCU	FU	Cal	%Fat	P	F	C	Na
Per Serving	0	0	138	6	1	1	34	4

Raspberry Apple Jam

1 pound Golden Delicious apples, cored and diced
6 tablespoons frozen unsweetened apple juice concentrate
6 tablespoons water
2 tablespoons frozen unsweetened pineapple juice concentrate
¾ cup chopped dried apples
1 pound fresh raspberries

1. Combine fresh apples, apple juice concentrate, water, and pineapple juice concentrate into heavy saucepan. Cover and cook over low heat 15 minutes.
2. Add dried apples and cook, uncovered, for 5 minutes.
3. Stir in raspberries, cover, cook over low heat for 5 minutes.
4. Remove cover and cook over medium heat for 10 to 15 minutes or until thickened, stirring frequently.
5. Quickly pour jam into hot jars, leaving ¼ inch headspace; cover at once with metal lids, and screw on bands. Process in boiling water bath 10 minutes. Or keep refrigerated.

Yield: 3 cups or 48-1 tablespoon servings

	RCU	FU	Cal	%Fat	P	F	C	Na
Per Serving	0	0	19	5	T	T	5	2

Fruit Pancake Topping

2 cups unsweetened applesauce
½ cup unsweetened crushed pineapple, drained
½ teaspoon cinnamon
1 8-ounce container plain no-fat or low-fat yogurt

1. Combine applesauce, pineapple, and cinnamon in a bowl. Chill.
2. Add yogurt before serving.
3. Serve on pancakes, waffles, french toast, etc.

Yield: 3½ cups topping or 14-¼ cup servings

	RCU	FU	Cal	%Fat	P	F	C	Na
Per Serving	0	0	30	0	1	T	7	13

Banana Jam

1 6-ounce can frozen pineapple juice OR apple juice concentrate
2 tablespoons cornstarch
3 cups sliced rice bananas
3 tablespoons lemon juice

1. Combine thawed juice concentrate and cornstarch in saucepan; stir until smooth. Mash bananas.
2. Add remaining ingredients, stirring well. Cook over medium heat, stirring constantly, until thickened.
3. Cool. Store in refrigerator.

Yield: 2 cups or 32-1 tablespoon servings

	RCU	FU	Cal	%Fat	P	F	C	Na
Per Serving	0	0	41	2	1	T	11	T

Strawberry Spread

2 envelopes unflavored gelatin
1 cup frozen unsweetened apple juice concentrate
2 quarts fresh strawberries (washed and hulled), mashed
1 tablespoon lemon juice

1. Sprinkle gelatin over ½ cup apple juice concentrate; set aside.

2. Combine remaining ½ cup apple juice concentrate, strawberries, and lemon juice in heavy saucepan. Cook over medium-low heat 10 to 15 minutes, stirring constantly.
3. Remove from heat and add softened gelatin mixture; stir until gelatin dissolves.
4. Cool and store in refrigerator.

Yield: 6 cups or 96-1 tablespoon servings

	RCU	FU	Cal	%Fat	P	F	C	Na
Per Serving	0	0	9	0	T	T	2	1

Fruit Yogurt

1 cup sliced banana
¼ cup crushed unsweetened pineapple,drained
1 teaspoon vanilla extract
½ teaspoon lemon juice
1 8-ounce container no-fat or low-fat plain yogurt
1 cup sliced strawberries

1. Combine banana, pineapple, vanilla and lemon juice in blender and process until smooth. Pour into a bowl.
2. Stir in yogurt and strawberries, blend well. Chill.

Yield: 4 servings

	RCU	FU	Cal	%Fat	P	F	C	Na
Per Serving	0	0	102	3	4	T	22	45

Breakfast Shake

1 8-ounce container plain no-fat or low-fat yogurt
1 large egg white
1 medium banana, sliced
¼ cup skim milk
½ cup unsweetened crushed pineapple, drained
¼ cup frozen unsweetenedpineapple juice concentrate
1 teaspoon vanilla extract

1. In blender, combine all ingredients. Blend at high speed for 30 seconds or until the mixture is smooth.

Yield: 4 (½ cup servings)

	RCU	FU	Cal	%Fat	P	F	C	Na
Per Serving	0	0	132	3	5	T	28	65

Granola

4 cups rolled oats
4 cups rolled wheat
2 cups wheat germ
1 cup unroasted sunflower seeds, optional
1 6-ounce can apple juice concentrate
2 tsp vanilla
¾ cup chopped dates
¾ cup raisins
¾ cup dried apples, cut into bite-size pieces
1 to 2 cups slivered almonds, optional

1. Mix together oats, rolled wheat, and wheat germ. Add nuts if desired.
2. Combine apple juice and vanilla.
3. Blend all ingredients together and put into a 9 x 13 cake pan.
4. Bake at 275 degrees for one hour.
5. After taking granola out of the oven, add the dried apples, dates and raisins.
6. Use as a cold cereal or mix with plain yogurt and fruit for a delicious breakfast.

Yield: 12 cups or 16-3/4 cup servings

	RCU	FU	Cal	%Fat	P	F	C	Na
Per Serving	0	0	275	10	10	3	56	7

Apple Carrot Muffins

½ cup skim milk
1 6-ounce can frozen unsweetened apple juice concentrate
3 large egg whites
1 teaspoon vanilla extract
2½ cups whole wheat flour
3 teaspoons baking powder
¼ teaspoon nutmeg
1 teaspoon cinnamon
1 cup grated apple
1 cup finely grated carrot
1 cup chopped nuts, optional

1. Combine milk, apple juice, egg whites, and vanilla; beat well.
2. In separate bowl, combine flour, baking powder, nutmeg, and cinnamon.
3. Add milk mixture to flour mixture; blend until just moistened. Stir in apple, carrot, and nuts.
4. Spray muffin tins with nonstick vegetable coating. Fill tins three-fourths full.
5. Bake at 400 degrees for 15 to 20 minutes.

Yield: 18 muffins

	RCU	FU	Cal	%Fat	P	F	C	Na
Per Muffin	0	0	87	4	3	0	19	67

Oatmeal Muffins

1 cup quick cooking oats
1 cup whole wheat flour
1 teaspoon baking powder
½ teaspoon baking soda
1 large egg white
½ teaspoon vanilla extract
½ teaspoon coconut extract
1 6-ounce can frozen unsweetened pineapple juice concentrate
½ cup unsweetened crushed pineapple, drained
½ cup buttermilk

1. Combine dry ingredients. Stir well.
2. Combine the remaining ingredients and add them to the dry
   ingredients. Stir until just moistened.
3. Coat muffin pans with nonstick cooking spray and distribute
   batter.
4. Bake at 400 degrees for 15 to 20 minutes or until brown.

Yield: 12 muffins

	RCU	FU	Cal	%Fat	P	F	C	Na
Per Muffin	0	0	107	3	3	T	30	76

Applesauce Cake

¼ cup cold-pressed oil
12-ounce frozen unsweetened apple juice concentrate
3 large egg whites
¼ cup unsweetened applesauce
1 teaspoon vanilla extract
2½ cup whole wheat flour
1 cup quick-cooking oats
¾ teaspoon baking soda
2 teaspoons baking powder
1 teaspoon cinnamon
½ teaspoon nutmeg
½ cup raisins
1 cup grated apple (1 large apple)

1. Preheat oven to 375 degrees. Spray 9 x 13 cake pan with nonstick vegetable coating. If you are going to remove the cake whole from the pan, line the bottom of the pan with waxed paper.
2. Beat together oil, apple concentrate, and egg whites.
3. Add applesauce and vanilla and combine well.
4. Mix together flour, oats, soda, baking powder, cinnamon, and nutmeg. Add to the liquid mixture and beat.
5. Add raisins and grated apple and stir until blended.
6. Pour batter into prepared pan.
7. Bake for 25 to 30 minutes or until a wooden pick inserted in the center comes out clean.

Yield: 24 servings

	RCU	FU	Cal	%Fat	P	F	C	Na
Per Serving	0	0	121	19	3	3	22	60

Date Cake

3 large egg whites
3 tablespoons cold-pressed oil
1 teaspoon vanilla extract
½ cup frozen unsweetened apple juice concentrate
½ cup unsweetenedapplesauce
½ cup banana, mashed
1 cup dates, chopped
2 cups whole wheat flour
⅓ cup wheat germ
2 teaspoons baking powder
1 teaspoon baking soda

1. Preheat oven to 350 degrees. Spray 9 x 9 baking pan with nonstick vegetable coating.
2. Combine egg whites and oil; beat well. Add vanilla, apple concentrate, applesauce and bananas; mix well.
3. Coat chopped dates in flour.
4. Combine flour-date mix, wheat germ, baking powder, and baking soda. Add to the liquid mixture. Mix well.
5. Pour into prepared pan. Bake for 25 to 30 minutes or until a wooden pick inserted into the center comes out clean.

Optional Topping:

Sprinkle chopped walnuts and additional chopped date pieces on top of the batter before it is baked. Bake as directed.

Yield: 16 servings

	RCU	FU	Cal	%Fat	P	F	C	Na
Per Serving	0	0	139	21	4	3	26	101

Banana-Oatmeal Cookies

2 cups whole wheat flour
¾ teaspoon baking soda
1 teaspoon cinnamon
½ teaspoon nutmeg
¼ cup cold-pressed oil
½ cup frozen unsweetened pineapple juice concentrate
2 egg large whites
1 cup (about three) mashed ripe bananas
1 teaspoon vanilla extract
2 cups rolled oats
½ cup chopped nuts (walnuts or pecans), optional

1. Combine oil, juice, egg whites, mashed bananas, and vanilla. Beat well.
2. Add flour, baking soda, cinnamon, and nutmeg. Mix well.
3. Stir in rolled oats and chopped nuts.
4. Drop by rounded teaspoonfuls onto cookie sheets that have been sprayed with nonstick vegetable coating.
5. Bake at 350 degrees for 10 to 12 minutes. When cookies are cool, place in a plastic bag. These cookies are better eaten the next day; banana flavor ripens overnight.

Yield: 2½ dozen cookies

	RCU	FU	Cal	%Fat	P	F	C	Na
Per Cookie	0	0	81	22	2	2	13	25

Apple Cookies

¼ cup cold-pressed oil
2 large egg whites
1 6-ounce can frozen unsweetened apple juice concentrate
1½ cups whole wheat flour
1½ cups quick cooking oats
1 teaspoon baking powder
1 teaspoon cinnamon
¼ teaspoon nutmeg
1 medium apple, peeled, cored, and finely diced
¼ cup raisins

1. Combine oil, egg whites, and apple juice; beat well.
2. Combine flour, oats, baking powder, cinnamon, and nutmeg; add to creamed mixture, mixing well.
3. Stir in diced apple and raisins.
4. Drop dough by rounded teaspoonfuls onto cookie sheets coated with nonstick vegetable coating.
5. Bake at 350 degrees for 10 to 12 minutes.

Yield: 2½ dozen cookies.

	RCU	FU	Cal	%Fat	P	F	C	Na
Per Cookie	0	0	71	28	2	2	12	15

BIBLIOGRAPHY

Chapter One

1. Merkel AD, Wayner MJ, Jolicoeur FB, Mintz RB. Effects of glucose and saccharin solutions on subsequent food consumption. Physiol Behav 1979;23:791-3.

2. Rolls BJ, Wood RJ, Stevens RM. Palatability and body fluid homeostasis. Physiol Behav 1978; 20:15-9.

3. Rolls BJ. Sweetness and Satiety. In: Sweetness. Ed: Dobbing J. Berlin: Springer-Verlag, 1987; pp. 161-73.

4. Tordoff MG. Sweet drinks increase food intake and preference in rats. Report at Symposium on Mechanisms of Appetite and Obesity. San Antonio, Texas: October 1985.

5. Lennon HD, Metcalf LE, et al. The biological properties of aspartame IV. Effect of reproduction and lactation. J Environmental Pathol Toxicol, 1980; 3:375-86.

6. Ishii H. Chronic feeding studies with aspartame and its diketopiperazine. In: Aspartame Physiology and Biochemistry. Eds: Stegink LD, Filer LJ. Jr. New York, NY: Marcel Dekker, 1984; pp. 307-19.

7. Braitman LE, Adlin EV, Stanton JL Jr. Obesity and caloric intake: The national health and nutrition examination survey of 1971-75 (HANES 1). J Chron Dis 1985; 38:727-32.

8. Garrow JS. Energy Balance and Obesity in Man. Elsevier, New York. 1974; pp. 84-85.

9. Wooley SC, Wooley OW, Dyrenforth SR. Theoretical, practical and social issues in behavioral treatments of obesity. J Appl Behav Analy 1979; 12:3-25.

10. Baoecke JAH, Van Staveren WA, Burema J. Food consumption, habitual physical activity, and body fatness in young Dutch adults. Am J Clin Nutr 1983; 37:295-99.

11. Keen H, Thomas BJ, Jarrett RJ, Fuller JH. Nutrient intake, adiposity and diabetes. Br Med J 1979;1: 655-58.

12. Kromhout D. Energy and macronutrient intake in lean and obese middle-aged men (the Zutphen Study). Am J Clin Nutr 1983; 37:295-99.

13. Johnson ML, Burke BS, Mayer J. The prevalence and incidence of obesity in a cross-section of elementary school children. Am J Clin Nutr 1956; 4:231-38.

14. Stefanik PA, Heald FP Jr, Mayer J. Caloric intake in relation to energy output of obese and non-obese adolescent boys. Am J Clin Nutr 1959; 7:55-62.

15. Keys A, Brozek J, et al. The Biology of Human Starvation. Minneapolis, MN: University of Minnesota Press, 1950; pp. 819-918.

16. Sims EAH. Studies in human hyperphagia. In: Treatment and Management of Obesity. Eds: Bray G, Bethune J. New York, NY: Harper and Row, 1974; pp. 29.

17. Sims EAH, Horton ES. Endocrine and metabolic adaptation to obesity and starvation. Amer J Clin Nutr 1968; 21:1455-70.

18. Sims EAH, Danforth EH Jr, et al. Endocrine and metabolic effects of experimental obesity in man. Recent Prog Horm Res 1973; 29:457-87.

19. Neuman RO. Experimentelle Beitrage zur Lehre von dem Taglichen Nahrungsbedarf des Menschen unter besonderer Berucksichtigung der notwendigen Eiwiessmenge. Archiv fur Hygiene 1902; 45:1-87.

20. Leveille GA. Adipose tissue metabolism: Influence of periodicity of eating and diet composition. Fed Proc 1970; 29:1294-1301.

21. Hollifield G, Parson W. Metabolic adaptations to a "stuff and starve" feeding program. I. Studies of adipose tissue and liver glycogen in rats limited to a short daily feeding period. J Clin Invest 1962; 41:245-59.

22. Bray G. The Obese Patient. Vol. IX, Major Problems In Internal Medicine. Ed: Smith LH. Jr. Philadelphia, PA: W.B. Saunders, 1976; pp. 176.

23. MacKay EM, Calloway JW, Barnes RH. Hyperalimentation in normal animals produced by protamine insulin. J Nutr 1940; 20:59-66.

24. Hoebel BG, Teitelbaum P. Weight regulation in normal and hypothalamic hyperphagic rats. J Comp Physiol Psychol 1066; 61:189-93.

25. Mayer J, Marshall NB, Vitale JJ, et al. Exercise, food intake, and body weight in normal rats and genetically obese adult mice. Amer J Physiol 1954; 177:544-8.

26. Mayer J, Thomas D. Regulation of food choice and obesity. Science 1967; 156:328-37.

27. Oscai LB, Holloszy JO. Effects of weight changes produced by exercise, food restriction, or overeating on body composition. J Clin Invest 1969; 48:2124-8.

28. Granneman JG, Wade GN. Effect of sucrose overfeeding on brown adipose tissue lipogenesis and lipoprotein lipase activity in rats. Metabolism 1983; 32:202-07.

29. Oscai LB, Spirakis CN, Wolff CA, et al. Effects of exercise and of food restriction on adipose tissue cellularity. J Lipid Res 1972; 13:588-92.

30. Castonguay TW, Hirsch E, Collier G. Palatability of sugar solutions and dietary selection. Physiol Behav 1981; 27:7-12.

31. Hirsch E, Dubois C, Jacobs HL. Overeating, dietary section patterns, and sucrose intake in growing rats. Physiol Behav 1982; 8.

32. Kanarek RB, Hirsch E. Developmental aspects of sucrose-induced obesity in rats. Physiol Behav 1979; 23:881-85.19-28.

33. Sclafani A, Xenakis S. Sucrose and polysaccharide induced obesity in the rat. Physiol Behav 1984; 32:169-74.

34. Schemmel R, Mickelsen O, Gill JL. Dietary obesity in rats: body weight and body fat accretion in seven strains of rats. J Nutr 1970; 100:1041-48.

35. Wade GN. Obesity without overeating in golden hamsters. Physiol Behav 1982; 29:701-07.

36. Oscai LB, Brown MM, Miller WC. Effect of dietary fat on food intake, growth and body composition in rats. Growth 1984; 48:415-24.

37. Greenwood MRC, Cleary M, et al. Adipose tissue metabolism and genetic obesity; the LPL hypothesis. In: Recent Advances in Obesity Research III. Eds: Bjorntrop P, Cairella M, Howard AN. London, Eng: John Libbey, 1981; pp. 75-79.

38. Powley TL. The ventromedial hypothalamic syndrome, satiety, and a cephalic phase hypothesis. Psychol Review 1977; 84/1:89-326.

39. Han PW. Hypothalamic obesity in rats without hyperphagia. Trans NY Acad Sci 1967; 30:229-43.

40. Love A, Hustvedt BE. Correlation between altered acetate utilization and hyperphagia in rats with ventromedial hypothalamic lesions. Metabolism 1973; 22:1459-65.

41. Jagot SA, Webb GP, et al. The induction of obesity in the rat with bipiperidly mustard. Br J Nutr 1980; 44:253-55.

42. Laughton W, Powley TL. Bipiperidyl mustard produces brain lesions and obesity in the rat. Brain Res 1981; 221:415-20.

43. Berthoud HR, Powley TL. Altered plasma insulin and glucose after obesity-producing bipiperidyl brain lesions. Am J Physiol 1985; 248:R46-53.

44. Martin WR, Wikier R, Eades CG, Pescor FT. Tolerance to and physical dependence on morphine in rats. Psychopharmacologia 1963; 4:247-60.

45. Bray GA. Obesity - a disease of nutrient or energy balance? Nutrition Reviews 1987; 45(2):33-43.

46. American Cancer Society. Cancer Prevention Study II, An epidemiological study of lifestyles and environment. CPS II Newsletter. Spring, 1986; 4/1:3.

47. Porikos KP, Hesser MF, Van Itallie TB. Caloric regulation in normal-weight men maintained on a palatable diet of conventional foods. Physiol Behav, Pergamon Press, 1982; 29:293-300.

48. Stegink LD, Filer LJ Jr. Aspartame: Physiology and Biochemistry. New York, NY: Marcel Dekker, Inc., 1984.

49. McCann MB, Trulson MF, Stulb SC. Non-caloric sweeteners and weight reduction. J Am Diet Assoc 1956; 32:327-30.

50. Rosenman K. Benefits of saccharin: A review. Environ Res 1978; 15:70-81.

51. Stare FJ. Sugar and sugar substitutes in preventative medicine and nutrition. Nutr Metab 1975; 18:133-42.

52. Walker ARP. The relative risks of saccharin and sucrose ingestion. Am J Clin Nutr 1979; pp. 727-28.

53. Alexander MM. Have formula diets helped? J Am Diet Assoc 1962; 40:538.

54. Parham ES, Parham AR. Saccharin use and sugar intake by college students. J Am Diet Assoc 1980; 76:560-63.

55. Adams PHSO, Grady KE, et al. Weight loss: long term results in an ambulatory setting. J Am Diet Assoc 1983; 83:306-10.

56. Finer N. Sugar substitutes in the treatment of obesity and diabetes mellitus. Clin Nutr 1985; 4:207-14.

57. Parham ES, Forbes CE. Saccharin use and sugar intake by college students. J Am Diet Assoc 1980; 76:560-63.6

58. Consumers Association. Which way to slim? Portsmouth, Eng: Eyre and Spottiswoode, 1978; pp. 96

59. Market Facts, Inc. Medical Serviced Group. Professional assessment of the physiological and psychological benefits of saccharin. A report to the Calorie Control Council. Chicago, IL: Market Facts Inc, 1978.

60. Cohen BL. Relative risks of saccharin and calorie ingestion. Science 1978; 199:983.

61. Brook M. The use of artificial sweeteners in food products. J R Soc Health 1969; 3:140-2.

62. Bernardis LL, McEwen G, Kodis M. Body weight set point studies in weanling rats with dorsomedial hypothalamic lesions (DMNL Rats). Brain Research Bulletin 1986; 17:451-60.

63. Garrow J. Energy Balance and Obesity in Man. New York, Elsevier, 1974.

64. Hamilton CL. Problems of refeeding and starvation in the rat. Ann New york Acad Sci. 1969; 157:1004-17.

65. Levitsky DA, Faust I, Glassman M. The ingestion of food and the recovery of body weight following fasting in the naive rat. Physiol Behav 1976; 17:575-80.

66. Boyle PC, Storlien H, Keesey RE. Increased efficiency of food unilization following weight loss. Physiol Behav 1978; 21:261-4.

67. Szepesi B. A model of nutritionally induced overweight: Weight "rebound" following caloric restriction. In: Bray GA. Ed: Recent Advances in Obesity Research. London, Newman, Ltd. 1978: Vol. 2.

68. Tepperman H, Tepperman J. Adaptive hyperlipogenesis. Fed Proc 1964; 23:73-5.

69. Greenwood MRC, Cleary M, et al. Adipose tissue metabolism and genetic obesity: the LPL hypothesis. In: Recent Advances In Obesity Research: III. Eds: Bjorntorp P, Cairella M, Howard AN. London, Eng: John Libbey, 1981; pp. 75-79.

70. Schwartz RS, Brunzell JD. Adipose tissue lipoprotein lipase and obesity. In: Recent Advances In Obesity Research: III. Eds: Bjorntorp P, Cairella M, Howard AN. London, Eng: John Libbey, 1981; pp. 94-97.

71. Rohner-Jeanrenaud R, Bobbioni E, et al. Central nervous system regulation of insulin secretion. In: CNS Regulation of Carbohydrate Metabolism. Ed: Szabo AJ. New York, NY: Academic Press, 1983; pp. 93-220.

72. von Borstel RW. Metabolic and physiologic effects of sweeteners. Clin Nutri 1985; 4/6:217.

73. Raymond CA. Obesity many disorders;' causes sought in genes, neurochemistry, psychology. JAMA 1986: 256:2301.

74. Woods SC. Conditioned hypoglycemia and conditioned insulin secretion. In: Advances in Metabolic Disorders: CNS Regulation of Carbohydrate Metabolism. Ed: Szabo AJ, 1983; pp. 494.

75. Strubbe JA, van Wachem P. Insulin secretion by the transplanted neonatal pancreas during food intake in fasted and fed rats. Diabetologia 1981; 20:228-36.

76. Steffens AB. Influence of the oral cavity on insulin release in the rat. Am J Physiol 1976; 230/5:1411-15.

77. Rohner-Jeanrenaud F, Bobbioni E, et al. Central nervous system regulation of insulin secretion. In: CNS Regulation of Carbohydrate Metabolism. Ed: Szabo AJ. New York, NY: Academic Press, Inc, 1983; pp. 193-220.

78. von Borstel RW. Metabolic and physiologic effects of sweeteners. Clin Nutr 1985; 4/6:215-20.

79. Powley TL. The ventromedial hypothalamic syndrome, satiety, and a cephalic phase hypothesis. Psychol Review 1977; 84/1:90.

80. Berthoud HR, Bereiter DA, et al. Cephalic phase, reflex insulin secretion. Diabetologia 1981; 20:393.

81. Louis-Sylvestre J. Preabsorptive insulin release and hypoglycemia in rats. Am J Physiol 1976; 230/1:56-60.

82. Raymond CA. TObesity many disorders;' causes sought in genes, neurochemistry, psychology. JAMA. 1986; 256/17:2301-07.

83. Raymond CA. Experts hold hope for obesity treatments targeted to specific regulatory miscues. JAMA 1986; 256/17:2302-07.

84. Berthoud HR, Bereiter DA, et al. Cephalic phase, reflex insulin secretion. Diabetologia 1981; 30:393-401.

85. Valenstein ES, Weber ML. Potentiation of insulin coma by saccharin. J Comparative Physiol Psychol 1965; 60/3:443-46.

86. Trimble ER, Berthoud HR, et al. Importance of cholinergic innervation of the pancreas for glucose tolerance in the rat. Am J Physiol 1981; 241:E337-41.

87. Kun E, Horvath I. The influence of oral saccharin on blood sugar. Proc Soc Exper Biol Med 1947; 66:175-77.

88. Steffens AB. Influence of the oral cavity on insulin release in the rat. Am J Physiol 1976; 230/5:1411-15.

89. Jorgensen H. The influence of saccharin on the blood sugar. Act Phys Scandinav 1949; 20:33-37.

90. Althausen TL, Wever GK. Proc Soc Exp Biol NY 1937; 35:517.

91. Kun E, Horvath I. The influence of oral saccharin on blood sugar. Proc Soc Exp Biol Med 1947; 66:175-77.

92. Louis-Sylvestre J. Preabsorptive insulin release and hypoglycemia in rats. Am J Physiol 1976; 230:56-60.

93. Sclafani A. Feeding inhibition and death produced by glucose ingestion in the rat. Physiol Behav 1973; 11:595-601.

94. Deutsch R. Conditioned hypoglycemia: a mechanism for saccharin-induced sensitivity to insulin in the rat. J Comparative Physiol Psychol 1974; 86/2:350-58.

95. Nicolaides S. Sensory-neuroendocrine reflexes and their anticipatory and optimizing role on metabolism. In: The Chemical Senses and Nutrition. Eds: Dare M, Maller O. New York, NY: The Nutrition Foundation, 1977; pp. 123-43.

96. Louis-Sylvestre J. Preabsorptive insulin release and hypoglycemia in rats. Am J Physiol 1976; 230/1:59.

97. Grossman MI. Pancreatic secretion in the rat. Am J Physiol 1958; 194:535-39.

98. von Borstel RW. Metabolic and physiologic effects of sweeteners. Clin Nutr 1985; 4/6:218.

99. Hernandez L, Hoebel BG. Basic mechanisms of feeding and weight regulation. In: Obesity Ed: Stunkard AJ. Philadelphia, PA: W.B. Saunders, 1980; pp. 25-47.

100. Rodin J. Effects of food choice on amount of food eaten in a subsequent meal: implications for weight gain. In: Recent Advance in Obesity Research: IV. Eds: Hirsch J, Van Itallie TB. London, Eng: John Libbey, 1985; pp. 101-12.

101. Stellar E. Contributions of physiological psychology to the understanding of human obesity. In: Recent Advances in Obesity Research: IV. Eds: Hirsch J, Van Itallie TB. London, Eng: John Libbey, 1985; pp. 113-17.

102. Bergman F, Cohen E, Leiblich I. Biphasic effects of chronic saccharin intake on pain responses of healthy and diabetic rats of two genetically selected strains. Psychopharmacology 1984; 82:248-51.

103. Lieblich I, Cohen E, et al. Morphine tolerance in genetically selected rats induced by chronically elevated saccharin intake. Science 1983; 221:871-3.

104. Siviy SM, Calcagnetti DC, Reid LD. Opioids and palatability In: The neural basis of feeding and reward. Eds: Hoebel BG, Novin R. Brunswick, Maine: Haer Institute for Electrophysiological Research, 1982; pp. 517-24.

105. Marks-Kaufman R, Kanarek R. Morphine selectively influences macronutrient intake in the rat. Pharmacol Biochem Behav 1980; 12:427-30.

106. Marks-Kaufman R, Lipeles BJ. Patterns of nutrient selection in rats orally self-administering morphine. Nutr Behav 1982; 1:33-46.

107. Atkinson RL. Opioid regulation of food intake and body weight in humans. Fed Proceed 1987; 46/1:178-82.

108. Faris PL, Komisaruk BR, et al. Evidence for the neuropeptide cholecystokinin as an antagonist of opiate analgesia. Science 1983; 219:310-12.

109. Genazzani AR, Facchinette F, et al. Hyperendorphinemia in obese children and adolescents. J Clin Endocrinol Metab 1986; 62:36-40.

Chapter Two

1. Lawlor PL. Sweet Talk - Media Coverage of Artificial Sweeteners. Washington D.C.: The Media Institute, 1986.

2. "America's Sweet Tooth," op. cit., pp. 50, and 1984 USDA estimates.

3. Morley JE, Levine AS, Yim GK, Lowy MT. Opioid modulation of appetite. Neurosci Biobehav Rev 1983; 7:281-305.

4. Atkinson RL. Opioid regulation of food intake and body weight in humans. Fed Proc 1987; 46:178-82.

5. Bergmann F, Cohen E, Lieblich I. Biphasic effects of chronic saccharin intake on pain response of healthy and diabetic rats of two genetically selected strains. Psychopharmacology 1984; 82:248-51.

6. Kral JG, Gortz L, Terenius L. Endorphin-like activity in morbidly obese patients. A pilot study. Int J Obes 1981; 5:539.

7. Herman CP, Polivy J. Restrained eating. In: Obesity. Ed: Stunkard AJ. Philadelphia, PA: W.B. Saunders, 1980; 208-25.

8. Reid RL, Yen SSC. Beta-endorphin stimulates the secretion of insulin and glucagon in humans. J Clin Endocrinol Metab 1981;52: 592-94.

9. Kanter RA, Ensinck JW, Fujimoto WY. Disparate effects of enkephalin and morphine upon insulin and glucagon secretion by islet cell culture. Diabetes 1980; 29:84-86.

10. Nowlis GH, Frank ME, Pfaffman C. Specificities of acquired aversions to taste qualities in hamsters and rats. J Comp Physiol Psychol 1980; 94:932-42.

11. Sato M, Hiji Y, Ito H, Imoto T. Sweet taste sensitivity in Japanese macaques. In: The Chemical Senses and Nutrition. Eds: Dare M. Maller O. New York, NY: The Nutrition Foundation, 1977; pp. 327-42.

12. Wurtman JJ. Neurotransmitter regulation of protein and carbohydrate consumption. In: Nutrition and Behavior. Ed: Miller SA. Philadelphia, PA: The Franklin Institute Press, 1981; pp. 69-75.

13. Stellar E. Contributions of physiological psychology to the understanding of human obesity. Recent Advances in Obesity Research: IV. Eds: Hirsch J, Van Itallie TB. London, Eng: John Libbey, 1985; pp. 14.

14. Jhanwar-Uniyal M, Leibowitz SF. Impact of food deprivation on alpha1 and alpha2 noradrenergic receptors in the paraventricular nucleus and other hypothalamic areas. Brain Research Bulletin 1986; 17:889-96.

15. Sjostrom L. A review of weight maintenance and weight changes in relation to energy metabolism and body composition. In: Recent Advances In Obesity Research: IV. Eds: Hirsch J, Van Itallie TB. London, Eng: John Libbey, 1985; pp. 82-94.

16. Jequier E, Schutz Y. Does a defect in energy metabolism contribute to human obesity? In: Recent Advances In Obesity Research: IV. Eds: Hirsch J, Van Itallie TB. London, Eng: John Libbey, 1985; pp. 76-81.

17. Buskirk ER. Analysis of the role of energy metabolism in the production of human obesity: an introduction based on observations from the satellite meeting on the regulation of energy expenditure. In: Recent Advances in Obesity Research: IV. Eds: Hirsch J, Van Itallie TB. Philadelphia, PA: John Libbey, 1985; pp. 71-81.

18. Jequier E. Does a thermogenic defect play a role in the pathogenesis of human obesity? Clin Physiol 1983; 3:1-7.

19. Raymond CA. 'Obesity many disorders;' causes sought in genes, neurochemistry, psychology. JAMA 1986; 256/17:2302.

20. Hernandez L, Hoebel BG. Basic mechanisms of feeding and weight regulation. In: Obesity. Ed: Stunkard AJ. Philadelphia, PA: W.B. Saunders Co., 1980; pp. 34,39.

21. Pardridge WM. Potential effects of dipeptide sweetener aspartame on the brain. In: Nutrition and the Brain Vol. 7. Eds: Wurtman RJ. Wurtman JJ. New York, NY: Raven Press, 1986; pp. 199-241.

22. Blundell JE, Hill AJ. Paradoxical effects of an intense sweetener (aspartame) on appetite. Lancet May 10, 1986; pp. 1092-93.

23. Sclafani A, Xenakis S. Sucrose and polysaccharide induced obesity in the rat. Physiol Behav, Pergamon Press Ltd., 1984; 32:169-74.

24. Blundell JE, Hill AJ. Analysis of hunger: inter-relationships with palatability, nutrient composition and eating. In: Recent Advances In Obesity Research: IV. Eds: Hirsch J, Van Itallie TB. London, Eng: John Libbey, 1985; pp. 118-29.

Chapter Three

1. Pritikin's arteries amazingly clean, pathologists say. American Medical News, March 22, 1985; pp. 29.

2. Cleave TL. The Saccharine Disease. Bristol, Eng: John Wright and Sons Ltd., 1974.

3. Ca-A Cancer Journal For Clinicians. Jan/Feb 1984; 34/1:18-19.

4. Higginson J. Summary: Nutrition and Cancer. Presented at the Workshop Conference on Nutrition in Cancer Causation and Prevention, October 18 to 20, 1982, Fort Lauderdale, FL. Cancer Research Supplement Vol. 43. No. 5 CNREA 8. pp. 2515s.

5. Select Subcommittee on Nutrition and Human Needs, United States Senate. Dietary Goals for the United States. Ed. 2, Stock No. 052-070-94376-8. Washington, D.C.: Government Printing Office, 1977.

6. Department of Health, Education, and Welfare. Healthy People: The Surgeon General's Report on Health Promotion and Disease Prevention. DHEW (Public Health Service) Publication No. 79-55071. Washington, D.C.: Government Printing Office, 1979.

7. American Medical Association, Council on Scientific Affairs. JAMA 1979; 242:2335-38.

8. Upton AC. Statement on Diet, Nutrition, and Cancer. Hearings of the Subcommittee on Nutrition, Senate Committee on Agriculture, Nutrition and Forestry, October 2, 1979.

9. United States Department of Agriculture and Department of Health, Education, and Welfare. Nutrition and Your Health - Dietary Guidelines for Americans. USDA-DHEW. Washington, D.C.: Government Printing Office, 1980.

10. National Academy of Sciences, Food and Nutrition Board. Toward Healthful Diets. Washington, D.C.: National Academy of Sciences, 1980.

11. American Heart Association, Committee on Nutrition. Rationale of the Diet-Heart Statement of the American Heart Association. Circulation, 65:839A-854A, 1982.

12. National Academy of Sciences, Committee on Diet, Nutrition, and Cancer. Diet, Nutrition, and Cancer. Washington, D.C.: National Academy Press, 1982.

13. National Board of Health and Welfare, Diet and Exercise. Report from the National Board of Health and Welfare, Stockholm, 1972.

14. Swedish National Food Administration. Swedish Nutrition Recommendations. Upsala, Sweden: The National Food Administration, 1981.

15. Royal Norwegian Ministry of Agriculture. On Norwegian Food and Nutrition Policy. Report No. 32(1975-1976) to the Storting, December, 1975.

16. Department of National Health and Welfare. Recommendations for Prevention Programs in Relation to Nutrition and Cardiovascular Disease. Ottawa, Canada. Bureau of Nutritional Sciences, Health Protection Branch, 1977.

17. Molitor GT. National nutrition goals - how far have we come? In: Critical Food Issues of the Eighties. Eds: Chou M, Harmon DP. Pergamon Press, Inc., 1979; pp. 135-41.

18. Langsford WA. A food and nutrition policy. Food Nutr Notes Rev, 1979; 36:100-103.

19. Kaw KT, Barret-Connor E. Dietary potassium, and stroke-associated mortality: A 12-year prospective population study. N Eng J Med 1987; 316:235-40.

20. Garcia-Palmieri MR, Sorlie P, et al. Relationship of dietary intake to subsequent coronary heart disease incidence: The Puerto Rico Heart Health Program. Am J Clin Nutr 1980; 33:1818-27.

21. Yano K, Rhoads GG, Kagan A, Tillotson J. Dietary intake and the risk of coronary heart disease in Japanese men living in Hawaii. Am J Clin Nutr 1978; 31:1270-79.

22. Kushi LH, Lew RA, Stare FJ, et al. Diet and 20-year mortality from coronary heart disease: The Ireland-Boston diet-heart study. N Eng J Med 1985; 312:811-18.

23. Morris N, Marr JW, Clayton DG. Diet and heart: a postscript. Brit Med J 1977; 2:1307-14.

24. Moore MC, Guzman MA, et al. Dietary-atherosclerosis study on deceased persons. J Am Diet Assoc 1976; 68:216-23.

25. Moore MC, Guzman MA, et al. Further data on the relation of selected nutrients to raised coronary lesions: Dietary-atherosclerosis study on deceased persons. J Am Diet Assoc 1977; 70:602-06.

26. Acheson RM, Williams DRR. Does consumption of fruit and vegetables protect against stroke? Lancet May 28, 1983; pp. 1191-93.

27. Wright A, Burstyn PG, Gibney MJ. Dietary fibre and blood pressure. Brit Med J 1979; 2:1541-43.

28. Khaw KT, Barret-Connor E. Dietary potassium and blood pressure in a population. Am J Clin Nutr 1984; 39:963-68.

29. Kromhout D, Bosschieter EB, Coulander CL. Potassium, calcium, alcohol intake and blood pressure: the Zutphen Study. Am J Clin Nutr 1985; 41:1299-1304.

30. Meneely GR, Battarbee HD. High sodium-low potassium environment and hypertension. Am J Cardiol 1976; 38:768-85.

31. McCarron DA. Is calcium more important than sodium in the pathogenesis of essential hypertension? Hypertension 1985; 7:607-27.

32. Reed D, McGee D, Yano K, Hankin J. Diet, blood pressure, and multicollinearity. Hypertension 1985; 7:405-410.

33. McCarron DA, Morris CD, et al. Blood pressure and nutrient intake in the United States. Science 1984; 224:1392-98.

34. Ernsberger PR. Neural mediation of genetic and nutritional effects on blood pressure: role of adrenergic receptor regulation in the kidney, brain, and heart. A dissertation submitted to the graduate school in partial fulfillment of the requirements of the degree Doctor of Philosophy. Northwestern University. August, 1984.

35. Brozek J, Chapman CB, Keys A. Drastic food restriction: effect on cardiovascular dynamics in normotensive and hypertensive conditions. JAMA 1948; 137:1569-74.

36. Harrison GF. Nutritional deficiency, painful feet, high blood pressure in Hong Kong. Lancet 1946; 1:961-64.

37. Stapleton R. Edema in recovered prisoners of war. Lancet 1946; 1:850-51.

38. Katsouyanni K, Trichopoulos D, et al. Diet and breast cancer: a case-control study in Greece. Int J Cancer 1986; 38:815-20.

39. Bjelke E. Epidemiologic studies of cancer of the stomach, colon and rectum, with special emphasis on the role of diet. Vol. 3. Ann Arbor, MI: University Microfilm, 1973; pp. 273-343.

40. Graham S, Dayal H. Diet in the epidemiology of cancer of the colon and rectum. JNCI 1978; 61:1709-14.

41. Colditz GA, Branch LG, et al. Increased green and yellow vegetable intake and lowered cancer deaths in an elderly population. Am J Clin Nutr 1985; 41:32-6.

42. Hennekens CH. Micronutrients and cancer prevention. N Eng J Med 1986; 315:1288-89.

43. Menkes MS, Comstock GW, et al. Serum beta-carotene, vitamins A and E, selenium, and the risk of lung cancer. N Eng J Med 1986; 315:1250-54.

44. Lipkin M, Newmark H. Effect of added dietary calcium on colonic epithelial-cell proliferation in subjecgts at high risk for familial colonic cancer. N Eng J Med 1985; 313:1381-84.

45. Merz B. Adding seeds to the diet may keep cancer at bay. JAMA 1983; 249/20:2746.

46. Raloff J. Coming-dietary aids to prevent cancer? Science News 1987; 131:206.

47. Ames BN. Dietary carcinogens and anticarcinogens. Science 1983; 221:1256-64.

48. Wong J. Cancer and chemicals..and vegetables. Chemtech July 1986; pp. 436-43.

49. Tannenbaum A. Nutrition and Cancer. New York: Hober-Harper, 1959; p.517.

50. Alcantara EN, Speckman EW. Diet, nutrition, and cancer. Amer J Clin Nutrition 1976; 29:1935-47.

51. Ross MH, Bras G. Lasting influence of early caloric restriction on prevalence of neoplasms in the rat. J Natl Cancer Inst. 1971; 47:1095-1113.

52. Strouthes A. Saccharin drinking and mortality in rats. Physiol Behav 1973; 10:781-91.

53. Sclafani A. Dietary Obesity. In: Obesity. Ed: Stunkard AJ. Philadelphia, PA: W.B. Saunders Company, 1980; pp. 167-81.

54. Richter CP. Total self-regulatory functions in animals and human beings. Harvey Lect, 1943; 38:63-103.

55. Hamilton LW. Starvation induced by sucrose ingestion in the rat: partial protection by septal lesions. J Comp Physiol Psychol 1971; 77:59-69.

56. Sclafani A. Feeding inhibition and death produced by glucose ingestion in the rat. Physiol Behav 1973; 11:595-601.

57. Harriman AE. The effect of a preoperative preference for sugar over salt upon compensatory salt selection by adrenalectomized rats. J Nutr 1955; 57:271-76.

58. Davis CM. The self-selection of diet experiment: its significance for feeding in the home. Ohio State Med J 1938; 34:862-68.

59. Davis CM. Self selection of diet by newly weaned infants: an experimental study. Am J Dis Child 1928; 36:651-79.

60. Story M, Brown JE. Do young children instinctively know what to eat? N Eng J Med 1987; 316:103-06.

61. Lawless H. Sensory development in children: research in taste and olfaction. J Am Diet Assoc 1985; 85:577-85.

62. Desor JA, Maller O, Turner RE. Taste in acceptance of sugars by human infants. J Comp Physiol Psychol 1973; 84:496-501.

63. Cowart BJ. Development of taste perception in humans: sensitivity and preference throughout the life span. Psychol Bull 1981; 90:43-73.

64. Stellar E. Contributions of physiological psychology to the understanding of human obesity. In: Recent Advances in Obesity Research: IV. Eds: Hirsch J. Van Itallie TB. London, Eng: John Libbey, 1985; pp. 113-17.

65. Hunter warns: Beware of Aspartame. The Human Ecologist. Nos. 23 & 24, Fall-Winter 1983-84; pp. 13.

66. Wretlind A. World sugar production and usage in Europe. In: Sugars in Nutrition. Eds: Sipple HL, McNutt KW. New York, NY: Academic Pres, 1974; pp. 81-92.

67. Blundell JE, Hill AJ. Paradoxical effects of an intense sweetener (Aspartame) on appetite. Lancet, May 10, 1986; pp. 1092-93.

68. Committee for a Study on Saccharin and Food Safety Policy. Saccharin: technical assessment of risks and benefits. Washington, D.C.: Assembly of Life Sciences/Institute of Medicine/National Research Council/National Academy of Sciences, 1978.

69. Nesbitt RE. Determinants of food intake in obesity. Science 1968; 159:1254-55.

70. Underwood PJ, Belton E, Hulme P. Aversion to sucrose in obesity. Proc Nutr Soc 1973; 32:93a.

71. Herman CP, Polivy J. Restrained Eating. In: Obesity. Ed: Stunkard. Philadelphia, PA: W.B. Saunders Company, 1980.

72. Stellar E. Contributions of physiological psychology to the understanding of human obesity. In: Recent Advances in Obesity Research: IV. Eds: Hirsch J, Van Itallie TB. London, Eng: John Libbey, 1985; pp. 117.

73. Atkinson RL. Opioid regulation of food intake and body weight in humans. Fed Proceed 1987; 46/1:180.

Chapter Four

1. Foster DW. Diabetes Mellitus. In: Harrison's Principles of Internal Medicine. 9th edition. New York, NY: McGraw Hill Book Company, 1980; pp. 1753.

2. Benson AE, et al. Insulin autoimmunity as a cause of hypoglycemia. Arch Intern Med Dec. 1984; 144:2351-54.

3. Soeldner JS, et al. Insulin-dependent diabetes mellitus and autoimmunity: islet-cell autoantibodies, insulin autoantibodies, and beta-cell failure. Letter to the editor. N Eng J Med. 1985; 313/14: 894.

4. Flier JS, et al. Receptors, antireceptor antibodies and mechanisms of insulin resistance. N Eng J Med 1979; 300:413.

5. Offenbacher E, Pi-Sunyer X. Beneficial effect of chromium-rich yeast on glucose tolerance and blood lipids in elderly subjects. Diabetes 1980; 29:919-25.

6. Fields M, Ferretti RJ, Smith JC. Jr, Reiser S. Effect of copper deficiency on metabolism and mortality in rats fed sucrose or starch diets. J Nutr 1983; 113:1335-45.

7. Bray GA. The Obese Patient. Philadelphia, PA: W.B. Saunders Company, 1976; pp.258.

8. Steffens AB. Influence of the oral cavity on insulin release in the rat. J Physiol May 1976; 230/5:1411-15.

9. Trimble ER, et al. Importance of cholinergic innervation of the pancreas for glucose tolerance in the rat. Am J Physiol 1981; 241:E337-341.

10. Perley MJ, Kipnis DM. Plasma insulin responses to oral and intravenous glucose: studies in normal and diabetic subjects. J Clin Investigation 1967; 46/12:1954-62.

11. Mahler RJ, Szabo O. Amelioration of insulin resistance in obese mice. Am J Physiol 1971; 221:980-83.

12. Mahler RJ. The pathogenesis of pancreatic islet cell hyperplasia and insulin insensitivity in obesity. In: Advances in metabolic disorders, Vol. 7. Eds: Levine R. Luft R. New York, NY: Academic Press, 1974; pp. 213-41.

13. Mahler RJ. The relationship between the hyperplastic pancreatic islet and insulin insensitivity in obesity. Acta Diabetol Lat 1981; 18:1-17.

14. Crapo PA. Dietary modifications in the management of diabetes. In: Handbook of Diabetes Vol. 5, Current and future therapies. Ed: Brownlee M. New York, NY: Garland STPM Press, 1981; pp. 46.

15. Hadden DR. Food and diabetes: The dietary treatment of insulin-dependent and non-insulin-dependent diabetes. Clin Endocrinol Metab 1982; 11:503-54.

16. Finer N. Sugar substitutes in the treatment of obesity and diabetes mellitus. Clin Nutr 1985; 207-14.

17. Meyer TW, Anderson S, Brenner BM. Dietary protein intake and progressive glomerular sclerosis: the role of capillary hypertension and hyperperfusion in the progression of renal disease. Annals of Internal Medicine 1983; 98(part 2):832-38.

18. West KM, Kalbfleisch JM. Influence of nutritional factors on prevalence of diabetes. Diabetes 1971; 20:99-108.

19. Kiehm TG, Anderson JW, Ward K. Beneficial effects of a high carbohydrate, high fiber diet on hyperglycemic diabetic men. Am J Clin Nutr 1976; 29:895-99.

20. Trowell HC. Dietary-fiber hypothesis of the etiology of diabetes mellitus. Diabetes 1975; 24:762-65.

21. Anderson JW, Sieling B. HCF Diets: A professional guide to high-carbohydrate, high-fiber diets. Lexington, KY: University of Kentucky Diabetes Fund, 1979.

22. Anderson JW. Dietary fiber in diabetes. In: Medical Aspects of Dietary Fiber. Eds: Spiller GA, Kay R. Plenum, New York. In press.

23. Rabinowitch IM. Effects of the high carbohydrate-low calorie diet upon carbohydrate tolerance in diabetes mellitus. Can Med Assoc J 1935; 33:136-44.

24. Kempner W, Peschel RL, Schlayer C. Effect of rice diet on diabetes mellitus associated with vascular disease. Postgrad Med 1958; 24:359-71.

25. Singh I. Low-fat diet and therapeutic doses of insulin in diabetes mellitus. Lancet 1955; 1:422-25.

26. Anderson JW, Ward K. Long term effects of high carbohydrate, high fiber diets on glucose and lipid metabolism: A preliminary report on patients with diabetes. Diabetes Care 1978; 1:77-82.

27. Pritikin N, McGrady PM Jr. The Pritikin program for diet and exercise. New York, NY: Grosset and Dunlap, Inc., 1979.

28. Coulston AM, Hollerbeck CB, et al. Metabolic effects of added dietary sucrose in individuals with noninsulin-dependent diabetes mellitus (NIDDM). Metabolism 1985; 34:962-66.

29. Court JM. Diet in the management of diabetes: why have special diet foods? Med J Aust 1976; 1:841-43.

30. Farkas CS, Forbes CE. Do non-calorie sweeteners aid patients with diabetes to adhere to their diets? J Am Diet Assoc 1965; 46:482-84.

31. Horwitz DL. Aspartame use by persons with diabetes. In: Aspartame: Physiology and Biochemistry. Eds: Stegink LD, Filer LJ. New York, NY: Marcel Dekker, Inc., 1984; pp. 633-39.

32. Horwitz DL. Can aspartame meet our expectations? J Am Diet Assoc. 1983; 83:142-46.

33. Tourian A. Control of phenylalanine hydroxylase synthesis in tissue culture by serum and insulin. J Cell Physiol 1975; 87:15-23.

Chapter Five

1. Stegink LD, Filer LJ Jr, Baker GL. Effect of aspartame and sucrose loading in glutamate-susceptible subjects. Am J Clin Nutr 1981;34/9:1899-905.

2. Settipane GA. The restaurant syndromes. Arch Intern Med 1986; 146:2129-30.

3. Cochran JW, Cochran AH. Monosodium glutamate: The Chinese restaurant syndrome revisited. JAMA 1984; 252:899.

4. Stevenson DD, Simon RA. Sensitivity to ingested metabisulfites in asthma subjects. J Allergy Clin Immunol 1981; 68:26-32.

5. Settipane GA. Adverse reactions to sulfites in drugs and foods. J Am Acad Dermatol 1984; 10:1077-80.

6. Science News 8/17/1985; pp. 100.

7. Bush RK, Taylor SL, Busse W. A critical evaluation of clinical trials in reactions to sulfites. J Allergy Clin Immunol July 1986; 78/1(Pt. 2):191-202.

8. Science News 12/13/1986; pp. 374.

9. Committee for a Study on Saccharin and Food Safety Policy: Saccharin: Technical Assessment of Risks and Benefits (Report No. 1). Washington, DC, Assembly of Life Sciences/Institute of Medicine, National Research Council/National Academy of Sciences, 1978.

10. Nakanishi K, Hagiwara A, Shibata M, et al: Dose response of saccharin in induction of urinary bladder hyperplasia in Fischer 344 rats pretreated with N-butyl-n-(4-hydroxybutyl-nitrosamine. JNCI 1980; 65:1005-10.

11. Fukushima S, Arai M, et al. Differences in susceptibility to sodium saccharin among various strains of rats and other animal species. Gann Feb 1983; 74/1;8-20.

12. Fukushima S, Hagiwara A, et al. Promoting effects of various chemicals in rat urinary bladder carcinogenesis initiated by N-nitroso-N-butyl-(4-hydroxybutyl) amine. Food Chem Toxicol 1983; 21:59-68.

13. Sakata T, Hasegawa R, et al. Inhibition by aspirin of N-[4-(5-nitro-2-furyl)-2-thiazolyl] formamide initiation and sodium saccharin promotion of urinary bladder carcinogenesis in male F344 rats. Cancer Res Aug 1986; 46/8:3903-06.

14. Cohen SM, Arai M, Jacobs JB, et al. Promoting effect of saccharin and DL-tryptophan in urinary bladder carcinogenesis. Cancer Res 1979; 39:1207-17.

15. Nakanishi K, Fukushima S, et al: Organ-specific promoting effects of phenobarbital sodium and sodium saccharin in the induction of liver and urinary bladder tumors in male F344 rats. JNCI 1982; 68:497-500.

16. Cohn V. Artificial sweetener wins FDA approval. Washington Post, July 14, 1981.

17. FDA denies CNI petition; more clearances granted. Food Chem News, Dec. 1, 1986; pp.40-45.

18. Roberts HJ. A clinician's adventures in medicine: is aspartame (NutraSweet) safe? On Call (The official publication of the Palm Beach County Medical Society) Jan 1987; pp. 16-20.

19. Evaluation of consumer complaints related to aspartame use. MMWR 1984; 33:243-49.

20. Johns DR. Migraine provoked by aspartame. N Eng J Med 1986; 315:456.

21. Mansfield LE. Food allergy and adult migraine: double-blind and mediator confirmation of an allergy etiology. Ann Allergy 1985; 55:126-29.

22. Grant ECG. Food allergies and migraine. Lancet May 5, 1979; pp. 966-68.

23. Kirn T. Discussion, ideas abound in migraine research; consensus remains elusive. JAMA 1987; 257:9-12.

24. Kirn T. Migraine:many things to many patients. JAMA 1987; 257:12-13.

25. Brown M, Gibney M, Husband PR, Radcliffe M. Food allergy in polysymptomatic patients. The Practitioner 1981; 225:1651-54.

26. Finn R, Cohen HN. Food Allergy: fact or fiction. Lancet February 25, 1978; pp. 426-28.

27. Anderson JA. Non-immunologically-mediated food sensitivity. Nutrit Reviews 1984; 42:109-15.

28. Walton RGW. Seizures and mania after high intake of aspartame. Psychosomatics April 1986; pp. 218-19.

29. Wurtman RJ. Aspartame: possible effect on seizure susceptibility. Lancet May 9, 1985; pp. 1060.

30. Pinto JMB, Maher T. High dose aspartame lowers the seizure threshold to subcutaneous pentylenetetrazol in mice. Pharmacologist 1986; 28:201.

31. Personal communication with Dr. Richard Wurtman.

32. Wurtman RJ. Possible relationship between aspartame (NutraSweet) consumption, seizures, and other CNS abnormalities. Introductory comments presented to the Food and Drug Administration, April 21, 1986.

33. Stegink LD. Aspartame metabolism in humans: acute dosing studies. In :Aspartame Physiology and Biochemistry. Eds: Stegink LD, Filer, LJ Jr. New York, NY: Marcel Decker, 1984; pp. 509-53.

34. Stegink LD, Filer LJ Jr, Baker GL. Effect of aspartame and aspartate loading upon plasma and erythrocyte free amino acid levels in normal adult volunteers. J Nutr 1977; 107:1837-45.

35. Stegink LD, Filer LJ Jr, Baker BL. Plasma, erythrocyte and human milk levels of free amino acids in lactating women administered aspartame or lactose. J Nutr 1979; 109:2173-81.

36. Fernstrom JD. Effects of acute aspartame ingestion on large neutral amino acids and monoamines in rat brain. In: Aspartame Physiology and Biochemistry. Eds: Stegink LD, Filer LJ. New York, NY: Marcel Dekker, 1984; pp. 641-53.

37. Wurtman RJ. Neurochemical changes following high-dose aspartame with dietary carbohydrates. N Eng J Med 1986; 315:429-30.

38. Coulombe FA Jr, Raghubir PS. Neurobiochemical alterations induced by the artificial sweetener aspartame (NutraSweet). Toxicol Appl Pharmacol 1986; 83:79-85.

39. Fernstrom JD, Fernstrom MH, Gillis MA. Acute effect of aspartame on large neutral amino acids and monoamines in rat brain. Life Sci 1983; 32:1651-58.

40. Stegink LD. Aspartame metabolism in humans: acute dosing studies. In: Aspartame Physiology and Biochemistry, Eds: Stegink LD, Filer LJ Jr. New York, NY: Marcel Dekker, 1984; pp. 509-53.

41. Schildkraut JJ. The catecholamine hypothesis of affective disorders: a review of supporting evidence. Am J Psychiatry 1965; 122:509-22.

42. Bunney WE Jr, Davis JM. Norepinephrine in depressive reactions. A review. Arch Gen Psychiatry 1965; 13:483-94.

43. Hanin I, Frazer J, et al. Depression: implications of clinical studies for basic research. Fed Proc 1985; 44/1 (Pt.1):85-90.

44. Kreuger DW. The depressed patient. In: Year Book of Family Practice. Ed: Rakel RE. Chicago, IL: Year Book Medical Publishers, Inc. 1980; pp. 111.

45. Hartmann E, Greenwald D. Tryptophan and human sleep: an analysis of 43 studies. In: Progress in Tryptophan and Serotonin Research. Eds: Schlossberger HB, Kochen W, Linzen B, Steinhart H. New York, NY: Walter de Gruyter 1984; pp. 297-304.

46. Yogman MW, Zeisel S. Diet and sleep patterns in newborn infants. N Eng J Med. 1983; 309:1147-49.

47. Yogman MW, Zeisel SH, Roberts C. Dietary precursors of serotonin and newborn behavior. J Psychiatry Res 1983; 17:123-33.

48. Geyer MA, Segal DS. Shock-induced aggression: opposite effects of intraventricularly infused dopamine and norepinephrine. Behav Biol 1974; 10:99-104.

49. Gibbons JL, Barr GA, Bridger WH, Leibowitz SF. Manipulations of dietary tryptophan: effects on mouse killing and brain serotonin in the rat. Brain Res 1979; 169:139-53.

50. Treiser SL, Cascio CS, et al. Lithium increases serotonin release and decreases serotonin receptors in the hippocampus. Science 1981; 213:1529-31.

51. Krassner MB. Brain chemistry. C&EN August 29, 1983; pp. 22-33.

52. Kohsada M, Hiramatsu M, Mori A. Brain catecholamine concentrations and convulsions in El mice. In: Advances in Biochemical Psychopharmacology. Eds: Roberts PJ, Woodruff GN, Iversen LL. New York, NY: Raven Press, 1978.

53. Hutchinson JSM. Control of the endocrine hypothalamus. In: Endocrine Hypothalamus. Eds: Jeffcoate SL, Hutchinson JSM. New York, NY: Academic Press, 1978.

54. Krulich L. Central neurotransmitters and the secretion of prolactin, GH, LH, and TSH. In: Annual Review of Physiology Eds: Edelman JS, Schultz SG. Palo Alto, CA: Annual Reviews, Inc., 1979.

55. Wurtman RJ. Nutrients that modify brain function. Scien Am 1982; 246/4; pp. 50-59.

56. Weisburd S. Food for mind and mood. Science News April 7,1984; 125:216-18.

57. Crandall EA, Fernstrom JD. Acute changes in brain tryptophan and serotonin after carbohydrate or protein ingestion by diabetic rats. Diabetes 1980; 29:960-66.

58. Fernstrom JD, Wurtman RJ. Brain serotonin content: increase following ingestion of carbohydrate diet. Science 1971; 174:1023-25.

59. Kolata G. Food affects human behavior. 1982; 218:1209-10.

60. Cole J, Hartmann E, Brigham P. L-tryptophan: clinical studies. McLean Hosp J 1980; 5:37-71.

61. Hartmann E, Lindsley JG, Spinweber C. Chronic insomnia: effects of tryptophan, flurazepam, secobarbital, and placebo. Psychopharmacology (Berlin) 1983; 80:138-42.

62. Schneider-Helmert D, Spinweber CL. Evaluation of L-tryptophan for treatment of insomnia: a review. Naval Health Research Center Report #84-4, San Diego, Ca; 1984.

63. Walinder J, Skott A, et al. Potentiation of the antidepressant action of clomipramine by tryptophan. Arch Gen Psychol 1976; 33:1384-89.

64. Glassman AH, Platman SR. Potentiation of a monoamine oxidase inhibitor by tryptophan. J Psychiatry Res 1969; 7:83-8.

65. Coppen A, Shaw DM, Herzberg B, Maggs R. Potentiation of the antidepressant effect of a monoamine-oxidase inhibitor by tryptophan. Lancet 1963; i:79-81.

66. Thorson J, Rankin H, et al. The treatment of depression in general practice: a comparison of L-tryptophan, amitriptyline, and a combination of L-tryptophan and amitriptyline with placebo. Psychol Med 1982; 12:741-51.

67. Gibson CJ, Wurtman RJ. Physiological control of brain catechol synthesis by brain tyrosine concentration. Biochem Pharmacol 1977; 26:1137-42.

68. Melamed E, Hefti F, Wurtman RJ. Tyrosine administration increases striatal dopamine release in rats with partial nigrostriatal lesions. NY Acad Sci 1980; 77:4305-09.

69. Wurtman RJ, Larin F, Mostafapour S, Fernstrom JD. Brain catechol synthesis: control by brain tyrosine concentration. Science 1974; 185:183-84.

70. Gelenberg AJ, Gibson CJ, Wojcik JD. Neurotransmitter precursors for the treatment of depression. Psychopharmacol Bull 1982; 18:7-18.

71. Gelenberg AJ, Wojcik JD, Gibson CJ, Wurtman RJ. Tyrosine for the treatment of depression. In: Research Strategies for Assessing the Behavioral Effects of Foods and Nutrients, Eds: Leiberman HR, Wurtman RJ. Proceedings of a conference held at the Massachusetts Institute of Technology, Cambridge, MA.

72. von Praag HM. In search of the action mechanism of antidepressants, 5-HTP/tyrosine mixtures in depression. Neuropharmacology 1983; 22:433-40.

73. von Praag HM, Lemus C. Monoamine precursors in the treatment of psychiatric disorders. In: Nutrition and the Brain. Eds: Wurtman RJ, Wurtman JJ. New York, NY: Raven Press, New York, 1986; pp. 116-18.

74. Sved AF, Fernstrom JD, Wurtman RJ. Tyrosine administration reduces blood pressure and enhances brain norepinephrine release in spontaneously hypertensive rats. Proc Natl Acad Sci USA, 1979; 76:3511-14.

75. Pardridge WM. Potential effects of the dipeptide sweetener aspartame on the brain. In: Nutrition and The Brain 7, Eds: Wurtman RJ, Wurtman JJ. New York, NY: Raven Press, 1986; pp. 199-241.

76. Kirn T. Migraine: Many things to many patients.JAMA 1987; 257(1):12-13.

77. Keys A, Brozek J, et al. The Biology of Human Starvation. Vol. 1. Minneapolis, MN: University of Minnesota Press, 1950.

78. Keys A, Brozek J, et al. The Biology of Human Starvation. Vol. 1. Minneapolis, MN: University of Minnesota Press, 1950.

79. Gilger AP, Potts AM. Studies on the visual toxicity of methanol V. The role of acidosis in experimental methanol poisoning. Am J Ophthalmol 1955; 39:63-86.

80. Roe O. The ganglion cell of the retina in cases of methanol poisoning in human beings and experimental animals. Acta Ophthalmol 1948; 26:169-82.

81. Fink WH. The ocular pathology of methyl-alcohol poisoning. Am J Ophthalmol 1943; 26:802-15.

82. Bennett IL Jr, Cary FH, Mitchell GL Jr, Cooper MN. Acute methyl alcohol poisoning: A review based on experiences in an outbreak of 323 cases. Medicine 1953; 32:431-63.

83. Tephly TR, McMartin KE. Methanol metabolism and toxicity. In: Aspartame Physiology and Biochemistry. Eds: Stegink LD, Filer LJ Jr. New York, NY: Marcel Dekker, 1984; pp. 111-40.

84. Gonda A, Gault H, Churchill D, Hollomby D. Hemodialysis for methanol intoxication. Am J Med 1978; 64:749-57.

85. McMartin KE, Ambre JJ, Tephly RT. Methanol poisoning in humans: Role for formic acid accumulation in the metabolic acidosis Am J Med 1980; 68:414-18.

86. Buller F, Wood CA. Poisoning by wood alcohol: Cases of death and blindness from Columbian spirits and other methylated preparations. JAMA 1904; 43:1117,1132,1289-96.

87. Benton CD, Calhoun FP. The ocular effects of methyl alcohol poisoning: Report of a catastrophe involving 320 persons. Am J Ophthalmol 1953; 36:1677-85.

88. Council on Scientific Affairs: Aspartame: Review of safety issues. JAMA 1985; 254:400-02.

89. Roak-Roltz R, Leveille GA. Projected aspartame intake: Daily ingestion of aspartic acid, phenylalanine, and methanol. In:Aspartame Physiology and Biochemistry. Eds: Stegink LD, Filer LJ. New York, NY: Marcel Dekker, 1984; pp. 201-05.

90. Monte WC. Aspartame: Methanol and the public health. J App Nutr 1984; 36:42-52.

91. Lund ED, Kirkland CL, Shaw PE. Methanol, ethanol, and acetaldehyde contents of citrus products. J Agri Food Chem 1981; 29:361-66.

92. Monte WC. Aspartame: methanol and the public health. J App Nutr 1984; 36/1:51.

93. Gilger AP, Potts AM, Farkas IS. Studies on the visual toxicity of methanol IX. The effect of ethanol on methanol poisonings in the rhesus monkey. Am J Ophthalmol 1956; 42:244-52.

94. Gilger AP, Farkas IS, Potts AM. Studies on the visual toxicity of methanol X. Further observations on the ethanol therapy of acute ethanol poisonings in monkeys. Am J Ophthalmol 1959; 48:153-61.

95. Monte WC. Aspartame: methanol and the public health. J App Nutr 1984; 36/1:50.

96. Bogert's Nutrition and Physical Fitness Tenth edition. Philadelphia, PA: W.B. Saunders Company, 1979; pp. 201.

97. Levine SA. Antioxidant Biochemical Adaptation: doorways to the new science and medicine. Prepublication Preview, 1984.

98. Stoewsand GS, Babish JB, Wimberly HC. J Environ Pathol Toxicol 1979; 2:399-406.

99. Wong JL. Cancer and chemicals...and vegetables. Chemtech July 1986; pp. 436-43.

100. World Health Organization (WHO). Indoor air pollutants: exposure and health effects assessment, Working Group Report, Nordlingen, Euro reports and studies No. 78, WHO, Copenhagen, Denmark, 1983.

101. Berglund B, Lindvall J, Sundell J (eds.). Indoor Air, Proceedings of the 3rd International Conference on Indoor Air Quality and Climate, Vol 1-6. Stockholm, Sweden, 1984.

102. National Academy of Sciences (NAS). Indoor Pollutants. Washington, USA: National Academy Press, 1981.

103. Gilbert S. Hazards of the toxic office. Sci Digest August 1984; pp. 24.

104. Fanger PO, Valbjorn O. (eds.) Indoor Climate. The Danish Building Research Institute. Copenhagen, Denmark, 1979.

105. Molhave L, Bach B, Pedersen OF. Dose-response relation of volatile organic compounds in the sick building syndrome. Clin Ecol 1986:4/2:52-6.

106. Rapp D. Allergies and Your Family. New York, NY: Sterling Publications, 1980; pp. 93.

107. Sprague DE, Milam MJ. Concept of an environmental unit. In: Food Allergy and Intolerance. Eds: Brostoff J, Challacombe SJ. London, Eng: Balliere Tindell, 1987; pp. 949-60.

108. Kroker GF. Chronic Candidiasis and Allergy. In: Food Allergy and Intolerance. Eds. Brostoff J, Challacombe SJ. London, Eng: Bailliere Tindell, 1987.

109. Kapner EM. Are 'carefree' fabrics carefree? In: Clinical Ecology. Ed. Dickey LD. Springfield, IL: Charles C. Thomas, 1976. 110. 20th Century Illness. Canadian Broadcasting Corporation, 1985.

110. 20th Century Illness, Canadian Broadcasting Corporation, 1985.

111. Boysen MR, Solberg LA. Changes in the nasal mucosa of furniture workers. A pilot study. Scand J Work Environ Health 1982; 8:273-82.

112. Gerhardsson MR, Norell SE, Kiviranta HJ, Ahlbom A. Respiratory cancers in furniture workers. Br J Ind Med 1985: 42:403-05.

113. Odkvist LM, Edling C, Hellquist H. Changes in the nasal mucosa after occupational exposure to wood dust and formaldehyde. Clin Ecol 1986:4/1:40-2.

114. Perry TL, Berry K, Hansen S, Diamond S, Mok C. Regional distribution of amino acids in human brain obtained at autopsy. J Neurochem 1971: 18:513-19.

115. Pardridge WM. Potential effects of the dipeptide sweetener aspartame on the brain. In: Nutrition and the Brain 7. Eds. Wurtman RJ, Wurtman JJ. New York, NY: Raven Press, 1987; pp.206.

116. Olney JW. Brain damage and oral intake of certain amino acids. Adv Exp Bio Med 1976; 69:497-506.

117. Price MT, Olney JW, Lowry OH, Buchsbaum S. Uptake of exogenous glutamate and aspartate by circumventricular organs but not other regions of brain. J Neurochem 1981; 36:1774-80.

118. Olney JW, Ho OL. Brain damage in infant mice following oral intake of glutamate, aspartate or cysteine. Nature 1970; 227:609-11.

119. Okaniwa A, Hori M, et al. Histopathological study on effects of potassium aspartate on the hypothalamus of rats. J Toxicol Sci 1979; 4:31-46.

120. Olney JW, Labruyere J, De Gubareff T. Brain damage in mice from voluntary ingestion of glutamate and aspartate. Neurobehav Toxicol 1980; 2:125-29.

121. Finkelstein MW, Daabees TT, Stegink LD, Applebaum AE. Correlation of aspartate dose, plasma dicarboxylic amino acid concentration, and neuronal necrosis in infant mice. Toxicol 1983; 29:109-19.

122. Garattini S. Evaluation of the neurotoxic effects of glutamic acid. In: Nutrition and the Brain 4. Eds: Wurtman RJ, Wurtman JJ. New York, NY: Raven Press, 1979; 79-124.

123. Pradhan SN, Lynch JR Jr. Behavioral changes in adult rats treated with monosodium glutamate in the neonatal stage. Arch Int Pharmacodyn Ther 1972; 197:301-04.

124. Berry HK, Butcher RE. Biochemical and behavioral effects of administration of monosodium glutamate to the young rat. Soc Neurosci (Abstr, 3rd annual meeting) 1973; pp. 8.

125. Vorhees CV, Butcher RE, Brunner RI, et al. A developmental test battery for neurobehavioral toxicity in rats: A preliminary analysis using monosodium glutamate, calcium carrageenan, and hydroxyurea. Toxicol Appl Pharmacol 1979; 50:267-82.

126. Olney JW, Ho O-L, Rhee V, et al. Neurotoxic effects of glutamate. N Eng J Med 1973; 289:1374-75.

127. Nemeroff CB, Lipton MA, Kizer JS. Models of neuroendocrine regulation: The use of monosodium glutamate as an investigative tool. Devel Neurosce 1978; 1:102-09.

128. Lamperti A, Blaha G. The effects of neonatally-administered monosodium glutamate on the reproductive system of adult hamsters. Biol Reprod 1976; 14:362-69.

129. Stegink LD, Reynolds WA, Filer LJ. Jr. et al. Comparative metabolism of glutamate in the mouse and man. In Filer LJ Jr. Garattini S, Dare MR, Reynolds WA, Wurtman RJ. (eds): Glutamic Acid: Advances in Biochemistry and Physiology. Raven Press, New York 1979:85-102.

130. Stegink LD. Aspartate and glutamate metabolism. In: Aspartame Physiology and Biochemistry. Eds: Stegink LD, Filer LJ Jr. New York, NY: Marcell Dekker, 1984; pp. 47-76.

131. Stegink LD. Aspartame metabolism in humans: acute dosing studies. In: Aspartame Physiology and Biochemistry. Eds. Stegink LD. Filer LJ Jr. New York, NY: Marcell Dekker, 1984; pp. 509-53.

132. Guroff G. Inborn errors of amino acid metabolism in relation to diet. In: Nutrition and Behavior. Ed: Miller AS. Philadelphia, PA: The Franklin Institute Press, 1981; pp. 77.

133. Holzman NA, Batshas ML, Valle DL. Genetic aspects of human nutrition. In: Modern Nutrition in Health and Disease. Eds: Goodhart RS, Shils ME. Philadelphia, PA: Lea and Febiger, 1980; pp. 1202-07.

134. Stegink LD, Filer LJ. Jr, Baker GL, McDonnell JE. Effect of aspartame loading upon plasma and erythrocyte amino acid levels in phenylketonuric heterozygotes and normal adult subjects. J Nutr 1979; 109:708-17.

135. Levy HL, Waisbren SE. Effects of untreated maternal phenylketonuria and hyperphenylalaninemia on the fetus. N Eng J Med 1983; 309:1269-74.

136. Koch R, Blaskovics M. Four cases of hyperphenylalaninemia: Studies during pregnancy and of the offspring produced. J Inher Metab Dis 1982; 5:11-15.

137. McKean CM. The effects of high phenylalanine concentrations on serotonin and catecholamine metabolism in the human brain. Brain Res 1972; 47:469-76.

138. Halestrap AP. Inhibition of mitochondrial pyruvate transport by phenylpyruvate and a-ketoisocaproate. Biochem Biophys Acta 1974; 367:102-08.

139. Harper AE, Benevenga NJ, Wohlhueter RM. Effects of ingestion of disproportionate amounts of amino acids. Physiol Rev 1970; 50:439-46.

140. Pratt OE. Transport inhibition in the pathogenesis of phenylketonuria and other inherited metabolic diseases. J Inhyer Metab Dis Suppl 1982; 2:75-81.

141. Stegink LD, Filer LJ Jr, Baker GL. Effect of aspartame and aspartate loading upon plasma and erythrocyte free amino acid levels in normal adult volunteers. J Nutr 1977; 107:1837-45.

142. Vaughan DA, Womack M, McClain PE. Plasma free amino acid levels in human subjects after meals containing lactalbumin, heated lactalbumin or no protein. Am J Clin Nutr 1977; 30:1709-12.

143. Yokogoshi H, Wurtman RJ. Acute effects of oral or parenteral aspartame on catecholamine metabolism in various regions of rat brain. J Nutr 1986; 116:356-64.

144. Yokogoshi H, Roberts CH, Caballero B, Wurtman RJ. Effects of aspartame and glucose administration on brain and plasma levels of large neutral amino acids and brain 5-hydroxyindoles. Am J Clin Nutr 1984; 40:1-7.

145. Yokogoshi H, Wurtman RJ. Acute effects of oral or parenteral aspartame on catecholamine metabolism in various regions of rat brain. J Nutr 1986; 116:356-64.

146. Stegink LD, Filer LJ Jr, Baker GL, McDonnell JE. Effect of aspartame loading upon plasma and erythrocyte amino acid levels in phenylketonuric heterozygotes and normal adult subjects. J Nutr 1979; 109:708-17.

147. Stegink LD. Aspartame metabolism in humans: Acute dosing studies. In: Aspartame Physiology and Biochemistry. Eds. Stegink LD, Filer LJ Jr. New York, NY: Marcel Dekker, 1984; pp. 525.

148. Uribe M. Potential toxicity of a new sugar substitute in patients with liver disease. N Eng J Med 1982; 306:173-74.

149. Jones MR, Kopple JD, Swendseid ME. Phenylalanine metabolism in uremic and normal man. Kidney Int 1978; 14:169-79.

150. Pardridge WM. Potential effects of the dipeptide sweetener aspartame on the brain. In: Nutrition and the Brain 7. Eds: Wurtman RJ, Wurtman JJ. New York, NY: Raven Press, 1986; pp. 231.

151. Kang E, Paine RS. Elevation of plasma phenylalanine levels during pregnancies of women heterozygous for phenylketonuria. J Pediatr 1963; 63:283-89.

152. Ranney RE, Mares SE, et al. The phenylalanine and tyrosine content of maternal and fetal body fluids from rabbits fed aspartame. Toxicol Appl Pharmacol 1975; 32:339-46.

153. Brass CA, Isaacs CE, McChesney R, Greengard O. The effects of hyper-phenylalaninemia on fetal development: A new animal model of maternal phenylketonuria. Pediatr Res 1982; 16:388-94.

154. Young M. The accumulation of protein by the fetus. In: Fetal Physiology and Medicine. Eds: Beard RW, Nathaniels PW. London, Eng: W. B. Saunders, 1976; pp. 59-79.

155. Lines DR, Waisman HA. Placental transport of phenylalanine in the rat: Maternal and fetal metabolism. Proc Soc Exp Biol Med 1971; 136:790-93.

156. Wurtman RJ. Possible relationship between aspartame (NutraSweet) consumption, seizures, and other CNS abnormalities. Introductory comments presented to the Food and Drug Administration, April 21, 1986.

157. Pitkin RM. Aspartame ingestion during pregnancy. In: Aspartame Physiology and Biochemistry. Eds: Stegink LD, Filer LJ Jr. New York, NY: Marcel Dekker, 1984; pp. 561.

158. Koch R, Wenz EJ. Aspartame ingestion by phenylketonuric heterozygous and homozygous individuals. In: Aspartame Physiology and Biochemistry. Eds. Stegink LD, Filer LJ Jr. 1984; pp. 593-606.

159. Pardridge WM. Potential effects of the dipeptide sweetener aspartame on the brain. In: Nutrition and the Brain 7. Eds: Wurtman RJ, Wurtman JJ. New York, NY: Raven Press, 1986; pp. 225.

160. Harper AE. Phenylalanine metabolism. In: Aspartame Physiology and Biochemistry. Eds: Stegink LD, Filer LJ Jr. New York, NY: Marcel Dekker, 1984; pp. 93.

161. Pardridge WM. Potential effects of the dipeptide sweetener aspartame on the brain. In: Nutrition and the Brain 7. Eds: Wurtman RJ, Wurtman JJ. New York, NY: Raven Press, 1986; pp. 225.

162. Levy HL, Waisbren SE. Effects of untreated maternal phenylketonuria and hyperphenylalaninemia on the fetus. N Eng J Med 1983; 309:1269-74.

163. Pardridge WM. Potential effects of the dipeptide sweetener aspartame on the brain. In: Nutrition and the Brain 7. Eds: Wurtman RJ, Wurtman JJ. New York, NY: Raven Press, 1986; pp. 226.

164. Krause W, Halminski M, et al. Biochemical and neuropsychological effects of elevated plasma phenylalanine in patients with treated phenylketonuria. J Clin Invest 1985; 75:40-48.

165. Felig P, Marliss E, Cahill GF Jr. Plasma amino acid levels and insulin secretion in obesity. N Eng J Med 1969; 281:811-15.

166. Fajans SS, Floyd JC Jr, Knopf RF, Conn JW. Effect of amino acids and proteins on insulin secretion in man. Recent Prog Horm Res 1967; 23:617-26.

167. File E-75 (1974). SC-18862: 104-week toxicity study in the mouse, PT 984H73, submitted by G. D. Searle and Co. to the Food and Drug Administration, Hearing Clerk File, Administrative Record, Aspartame 75F-0355, Food and Drug Administration, Rockville, Md.

168. File E-87 (1975). SC-18862: A supplemental evaluation of rat brains from two tumorigenicity studies, PT 1227, submitted by G. D. Searle and Co. to the Food and Drug Administration, Hearing Clerk File, Administrative Record, Aspartame 75F-1355, Food and Drug Administration, Rockville, Md.

169. Ishii H. Chronic feeding studies with aspartame and its diketopiperazine. In: Aspartame Physiology and Biochemistry. Eds: Stegink LD, Filer LJ Jr. New York, NY: Marcel Dekker, 1984; pp. 307-19.

170. Ishii H. Chronic feeding studies with aspartame and its diketopiperazine. In: Aspartame Physiology and Biochemistry. Eds: Stegink LD, Filer LJ Jr. New York, NY: Marcel Dekker, 1984; pp. 307-19.

171. File E-28 (1973). SC-18862: 106-week oral toxicity study in the dog, submitted by G. D. Searle and Co. to the Food and Drug Administration, Hearing Clerk File, Administrative Record, Aspartame 75F-0355, Food and Drug Administration, Rockville, Md.

172. Butcher RE, Vorhees CV. Behavioral testing in rodents given food additives. In: Aspartame Physiology and Biochemistry. Eds: Stegink LD, Filer LJ Jr. New York, NY: Marcell Dekker, 1984; pp. 398.

173. Brunner RL, Vorhees CV, Kinney L, Butcher RE. Aspartame: assessment of developmental psychotoxicity in a new artificial sweetener. Neurobehav Toxicol 1979; 1:79-86.

174. Potts WJ, Bloss JL, Nutting EF. Biological properties of aspartame. 1. Evaluation of central nervous system effects. J Environ Pathol Toxicol 1980; 3:341-53.

175. Visek WJ. Chronic ingestion of aspartame in humans. In: Aspartame Physiology and Biochemistry. Eds: Stegink LD, Filer LJ Jr. New York, NY: Marcel Dekker, 1985; 500,505,506.

176. Koch R, Shaw KNF, Williamson M, Haber M. Use of aspartame in phenylketonuric heterozygous adults. J Toxicol Environ Health 1976; 2:453-57.

177. Stegink LD. Aspartame metabolism in humans: Acute dosing studies. In: Aspartame Physiology and Biochemistry. Eds: Stegink LD, Filer LJ Jr. New York, NY: Marcel Dekker, 1984; pp. 521.

178. Suomi SJ. Effects of aspartame on the learning test performance of young stumptail macaques. In: Aspartame Physiology and Biochemistry. Eds: Stegink LD, Filer LJ Jr. New York, NY: Marcel Dekker, 1984; pp. 443.

179. Visek WJ. Chronic ingestion of aspartame in humans. In: Aspartame Physiology and Biochemistry. Eds: Stegink LD, Filer LJ Jr. New York, NY: Marcel Dekker, 1984; pp. 507.

180. Pitkin RM. Aspartame ingestion during pregnancy. In: Aspartame Physiology and Biochemistry. Eds: Stegink LD, Filer LJ Jr. New York, NY: Marcel Dekker, 1984; pp. 561.

181. Molinary SV. Preclinical studies of aspartame in nonprimate animals. In: Aspartame Physiology and Biochemistry. Eds: Stegink LD, Filer LJ Jr. New York, NY: Marcel Dekker, 1984; pp. 299.

182. Roberts HJ. A clinician's adventures in medicine: is aspartame (NutraSweet) safe? On Call (The official publication of the Palm Beach County Medical Society). January, 1987; pp. 16.

Chapter Six

1. Alias AB. Increased production and consumption of cane sugar in the third world to prevent undernutrition and famine. N Eng J Med Oct 3, 1985; pp. 894.

2. Hallfrisch J, Cohen L, Reiser S. Effects of feeding rats sucrose in a high fat diet. J Nutr 1981; 111:531-36.

3. Reiser S, Michaelis IV O, Putney J, Hallfrisch J. Effect of sucrose feeding on the intestinal transport of sugars in two strains of rats. J Nutr 1975; 105:894-905.

4. Gardner LB, Spannhake EB, Keeney M. Effect of dietary carbohydrate on serum insulin and glucagon in two strains of rats. Nutr Rept Int 1977; 15:361-66.

5. Vrana A, Slabochova Z, Fabry P, Kazdova I. Influence of diet with a high starch or sucrose content on glucose tolerance, serum insulin level, and insulin sensitivity in rats. Physiol Bohemeslov 1974; 23:305-10.

6. Michaelis IV OE, Ellwood KC, Hallfrisch J, Hansen CT. Effect of dietary sucrose and genotype on metabolic parameters of a new strain of genetically obese rat: LA/N-Corpulent. Nutr Res 1983; 3:217-28.

7. Vrana A, Slabochova Z, Kazdova L, Fabry P. Insulin sensitivity of adipose tissue and serum insulin concentration in rats fed sucrose or starch diets. Nutr Rep Int 1971; 3:31-7.

8. Romsos DR, Leveille GA. Effect of meal frequency and diet composition on glucose tolerance in the rat. J Nutr 1974; 104:1503-12.

9. Cohen AM. Genetically determined response to different ingested carbohydrates in the production of diabetes. Horm Metab Res 1978; 10:86-92.

10. Laube H, Wojcikowski C, Schatz H, Pfeifer EF. The effect of high maltose and sucrose feeding on glucose tolerance. Horm Metab Res 1978; 10:192-95.

11. Reiser S, Handler HB, et al. Isocaloric exchange of dietary starch and sucrose in humans. II. Effect on fasting blood insulin, glucose, and glucagon and on insulin and glucose response to a sucrose load. Am J Clin Nutr 1979; 32:2206-16.

12. Reiser S, Bohn E, et al. Serum insulin and glucose in hyperinsulinemic subjects fed three different levels of sucrose. Am J Clin Nutr 1981; 34:2348-58.

13. Behall KM, Moser PB, Kelsay JL, Prather ES. The effect of kind of carbohydrate in the diet and use of oral contraceptives on metabolism of young women -III. Serum glucose, insulin, and glucagon. Am J Clin Nutr 1980; 33:1041-48.

14. Beck-Nielsen H, Pedersen O, Sorensen NS. Effects of diet on the cellular insulin binding and the insulin sensitivity in young healthy subjects. Diabetes 1978; 15:189-96.

15. Hallfrisch J, Lazar F, Jorgensen C, Reiser S. Insulin and glucose responses in rats fed sucrose or starch. J Nutr 1981; 32:787-93.

16. Reiser S, Hallfrisch J. Insulin sensitivity and adipose tissue weight of rats fed starch or sucrose diets ad libitum or in meals. J Nutr 1977; 107:147-55.

17. Michaelis IV OE, Ellwood KC, Judge JM, Schoene NW, Hansen CT. Effect of dietary sucrose on the SHR/N-corpulent rat: a new model for insulin-independent diabetes. Am J Clin Nutr 1984; 39:612-18.

18. Obell AE. Recent advances in mechanism of causation of diabetes mellitus in man and Acomys cahirinus. E Afr Med J 1974; 51:425-28.

19. Hallfrisch J, Lazar FL, Reiser S. Effect of feeding sucrose or starch to rats made diabetic with streptozotocin. J Nutr 1979; 109:1909-15.

20. Keen H, Thomas BJ, Jarrett RJ, Fuller JH. Nutrient intake, adiposity, and diabetes. Br Med J 1979; 1:655-58.

21. Olsen ME, Faber OK, Binder C. Hepatic extraction of insulin after carbohydrate hyperalimentation. Acta Endocrinol 1983; 102:416-19.

22. Hassinger W, Sauer G, et al. The effects of equal caloric amounts of xylitol, sucrose, and starch on insulin requirements and blood glucose levels in insulin-dependent diabetics. Diabetes 1981; 21:37-40.

23. Kiehm TG, Anderson JW, Ward K. Beneficial effects of a high carbohydrate, high fiber diet on hyperglycemic diabetic men. Am J Clin Nutr 1976; 29:895-99.

24. Anderson JW. High polysaccharide diet studies in patients with diabetes and vascular disease. Cereal Foods World 1977; 22:12-13,22.

25. Anderson JW, Ward K. Long-term effects of high-carbohydrate, high-fiber diets on glucose and lipid metabolism: a preliminary report on patients with diabetes. Diabet Care 1978; 1:77-82.

26. Simpson HCR, Lousley S, et al. A high carbohydrate leguminous fibre diet improves all aspects of diabetic control. Lancet 1981; 1:1-5.

27. Naughton JM, O'Dea K, Sinclair AJ. Animal foods in traditional Australian Aboriginal diets: polyunsaturated and low in fat. Lipids 1986; 21/11:684-90.

28. West KM, Kalbfleisch JM. Influence of nutritional factors on prevalence of diabetes. Diabetes 1971; 20;99-108.

29. Reiser S, Hallfrisch J, Lyon R, Michaelis IV OE. Effect of chronic hyperinsulinism on metabolic parameters and histopathology in rats fed sucrose or starch. J Nutr 1983; 107:1073-80.

30. Leiter EH, Coleman DL, Ingram DK, Reynolds MA. Influence of dietary carbohydrate on the induction of diabetes in C57BL/KsJ-db/db diabetes mice. J Nutr 1983; 113:184-95.

31. Reiser S, Hallfrisch J, Putney J, et al. Enhancement of intestinal sugar transport by rats fed sucrose as compared to starch. Nutr Metab 1976; 20:461-70.

32. Gray RS, Olefsky JM. Effect of a glucosidase inhibitor on the metabolic response of diabetic rats to a high carbohydrate diet, consisting of starch and sucrose, or glucose. Metabolism 1983; 31:88-92.

33. Schemmel RA, Teague FJ, Bray GA. Obesity in Osborne-Mendel and S 5B/PI rats: effects of sucrose solutions, castration, and treatment with estradiol or insulin. J Physiol 1982; pp. R347-53.

34. Muto S, Miyahara C. Eating behavior of young rats: experiments on selective feeding of diet and sugar solutions. Brit J Nutr 1972; 28:327-37.

35. Kratz C, Levitsky D. Dietary obesity: differential effects with self-selection and composite diet feeding techniques. Physiol Behav 1979; 22:245-49.

36. Clevidence BA, Srinivasan SR, et al. Serum lipoprotein and blood pressure levels in rhesus monkeys fed sucrose diets. Biochem Med 1981; 25:186-97.

37. Srinivasan SR, Berenson GS, et al. Effects of dietary sodium and sucrose on the induction of hypertension in spider monkeys. Am J Clin Nutr 1980; 33:561-69.

38. Granneman JG, Wade GN. Effect of sucrose overfeeding on brown adipose tissue lipogenesis and lipoprotein lipase activity in rats. Metabolism 1983; 32:202-07.

39. Sheehan PM. Blood glucose and plasma lipids of Zucker fatty and lean rats fed diets containing cornstarch and sucrose. Diss Abstr Int 1981; 42:2312B.

40. Kanarek RG, Orthen-Gambill N. Differential effects of sucrose, fructose and glucose on carbohydrate-induced obesity in rats. J Nutr 1982; 112:1546-54.

41. Kanarek RB, Hirsch E. Dietary-induced overeating in experimental animals. Fed Proc 1977; 36:154-58.

42. Granneman JG, Campbell RG. Effects of sucrose feedings and denervation on lipogenesis in brown adipose tissue. Metabolism 1984; 33:257-61.

43. Hill W, Castonguay TW, Collier GH. Taste or diet balancing? Physiol Behav 1980; 24:765-67.

44. Allen RJL, Leahy JS. Some effects of dietary dextrose, fructose, liquid glucose and sucrose in the adult male rat. Brit J Nutr 1966; 20:339-47.

45. Lock S, Ford MA, Bagley R, Green LF. The effect on plasma-lipids of the isoenergetic replacement of table sucrose by dried-glucose syrup (maize-syrup solids) in the normal diet of adult men-over a period of 1 year. Br J Nutr 1980; 43:251-56.

46. Rath R, Masek J, Kujalova V, Slabochova Z. Effect of a high sugar intake on some metabolic and regulatory indicators in young men. Nahrung 1974; 18:343-53.

47. Porikos KP, Van Itallie TB. Efficacy of low-calorie sweeteners in reducing food intake: studies with aspartame. In: Aspartame Physiology and Biochemistry. Eds: Stegink LD, Filer LF Jr. New York, NY: Marcel Dekker, 1984; pp. 273-86.

48. Olsen ME, Faber OK, Binder C. Hepatic extraction of insulin after carbohydrate hyperalimentation. Acta Endocrinol 1983; 102:416-19.

49. Sclafani A, Springer D. Dietary obesity in adult rats: similarities to hypothalamic and human obesity syndromes. Physiol Behav 1976; 17:461-71.

50. Sclarani A, Rendel A. Food deprivation-induced activity in dietary obese, dietary lean, and normal-weight rats. Behav Biol 1978; 24:220-28.

51. Rolls BJ, Rowe EA. Dietary obesity: permanent changes in body weight. J Physiol 1976; 272:2P.

52. Rothwell NJ, Stock MJ. Mechanisms of weight gain and loss in reversible obesity in the rat. J Physiol 1978; 276:60P-61P.

53. Rolls BJ. How variety and palatability can stimulate appetite. Nutr Bull 1979; 5:78-86.

54. Rolls BJ, Rolls ET, Rowe EA. The influence of variety on human food selection and intake. In: The Psychobiology of Human Food Selection. Ed: Baker LM. West Port, Conn. 1982; pp. 101-22.

55. Ahrens RA, Demuth P, Lee MK, Majkowski JW. Moderate sucrose ingestion and blood pressure in the rat. J Nutr 1980; 110:725-31.

56. Beebe CG, Schemmel R, Michelson O. Blood pressure of rats as affected by diet and concentration of NaCl in drinking water. Proc Soc Exp Biol Med 1976; 151:205-99.

57. Smith-Barbaro PA, Quinn MR, Fisher H, Hegsted DM. Pressor effects of fat and salt in rats. Proc Soc Exp Biol Med 1980; 165:283-90.

58. Michaelis IV OE, Martin RE, Gardner LB, Ellwood KC. Effect of dietary carbohydrates on systolic blood pressure of normotensive and hypertensive rats. Nutr Rept Inf 1981; 23:261-66.

59. Bunag RD, Tomita T, Sasaki S. Chronic sucrose ingestion induces mild hypertension and tachycardia in rats. Hypertension 1983; 5:218-25.

60. Hodges RE, Rebello T. Carbohydrates and blood pressure. Ann Intern Med 1983; 98:838-41.

61. Israel KD, Michaelis IV OE, Reiser S, Keeney M. Serum uric acid, inorganic phosphorus, and glutamic-oxalacetic transaminase and blood pressure in carbohydrate-sensitive adults consuming three different levels of sucrose. Ann Nutr Metab 1983; 27:425-35.

62. Preuss MB, Preuss HG. The effects of sucrose and sodium on blood pressures in various substrains of Wistar rats. Lab Invest 1980; 43:101-07.

63. Young JB, Lindsberg L. Effect of oral sucrose on blood pressure in the spontaneously hypertensive rat. Metabolism 1981; 30:421-24.

64. Hall CE, Hall O. Comparative ability of certain sugars and honey to enhance saline polydipsia and salt hypertension. Proc Soc Exp Biol Med 1966; 122:362-65.

65. Williams CA, MacDonald I. Metabolic effects produced in baboons associated with the ingestion of diets based on lactose hydrolysate. Ann Nutr Metab 1982; 26:374-83.

66. Waterman RA, Romsos DR, et al. Effects of dietary carbohydrate source on growth, plasma metabolites, and lipogenesis in rats, pigs and chicks. Proc Soc Exp Biol Med 1975; 150:220-25.

67. Vijayagopal P, Srinivasan SR, et al. Decreased secretion of triacylglycerol of exogenous cholesterol in high sucrose-fed rabbits. Biochem Med 1980; 24:49-59.

68. Sebastian KL, Zacharias NT, Philip LB, Augusti KT. The hypolipidemic effect of onion (Allium cepa Linn) in sucrose fed rabbits. Indian J Physiol Pharmacol 1979; 23:27-30.

69. Mattock MB, Sheorain VS, Subrahmanyam D. Lecithin-cholesterol acyltransferase activity in carbohydrate-induced hypertriglyceridemia in mice. An Immunofluorescent method for identification of isolated thyrotropic cells. Experientia 1978; 34:304-05.

70. Shiff TS, Roheim PS, Eder HA. Effects of high sucrose diets and 4-aminopyrazolopyrimidine on serum lipids and lipoproteins in the rat. J Lipid Res 1971; 12:596-603.

71. Holt PR, Dominguez AA, Kwartler J. Effect of sucrose feeding upon intestinal and hepatic lipid synthesis. Am J Clin Nutr 1979; 32:1792-98.

72. Bruckdorfer DR, Kang SS, et al. Diurnal changes in the concentrations of plasma lipids, sugars, insulin and corticosterone in rats fed diets containing various carbohydrates. Horm Metab Res 1974; 6:99-106.

73. Reaven GM, Risser TR, Chen Y-DI, Reaven EP. Characterization of a model of dietary-induced hypertriglyceridemia in young, nonobese rats. J Lipid Res 1979; 20:371-78.

74. Vijayagopalan P, Kurup PA. Effect of dietary starches on the serum, aorta and hepatic lipid levels in cholesterol-fed rats. Atherosclerosis 1970; 11:257-64.

75. Berdanier CD. Metabolic characteristics of the carbohydrate-sensitive BHE strain of rats. J Nutr 1974; 104:1246-56.

76. Sheorain VS, Mattock MB, Subrahmanyam D. Mechanisms of carbohydrate-induced hypertriglyceridemia: Plasma lipid metabolism in mice. Metabolism 1980; 29:924-29.

77. Kelly TJ, Holt PR, Wu A-L. Effect of sucrose on intestinal very low-density lipoprotein production. Am J Clin Nutr 1980; 33:1033-40.

78. MacDonald I, Grenby TH, Fisher MA, Williams C. Differences between sucrose and glucose diets in their effects on the rate of body weight change in rats. J Nutr 1981; 111:1543-47.

79. Naismith DJ, Rana IA. Sucrose and hyperlipidemia II. The relationship between the rates of digestion and absorption of different carbohydrates and their effects on enzymes of tissue lipogenesis. Nutr Metab 1974; 16;285-94.

80. Sheehan PM, Reynolds LR, Thye FW, Ritchey SJ. Blood glucose and plasma lipids of Zucker rats fed diets containing cornstarch or sucrose. Nutr Rept Int 1984; 20:1337-44.

81. Ross AC, Minick CR, Zilversmit DB. Equal atherosclerosis in rabbits fed cholesterol free low fat diet or cholesterol supplemented diet. Atherosclerosis 1978; 29:301-15.

82. Kritchevsky D, Tepper SA, Kitagawa M. Experimental atherosclerosis in rabbits fed cholesterol-free diets. 3. Comparison of fructose and lactose with other carbohydrates. Nutr Rep Int 1973; 7:193-202.

83. Seely S, Horrobin DR. Diet and breast cancer: the possible connection with sugar consumption. Med Hypoth 1983; 11:319-27.

84. Graham S, Marshall J, et al. Diet in the epidemiology of breast cancer. Am J Epidemiol 1982; 116:68-75.

85. Cancer Statistics 1987. Ca - A Cancer Journal for Clinicians. Published by the American Cancer Society. Jan/Feb 1987; 37/1:16-17.

86. Levin AS, Zellerbach M. The Type 1/Type 2 Allergy Relief Program. Los Angeles, CA: Jeremy P. Tarcher, Inc., 1983.

87. Food Allergy. Ed: Gerrard JW. Springfield, IL: Charles C. Thomas, 1980.

88. Mandell M, Scanlon LW. 5-Day Allergy Relief System. New York, NY: Pocket Books, 1980.

89. Philpott WH, Kalita DK. Brain Allergy. New Canaan, CT: Keats Publishing, 1980.

90. Mandell M, Mandell FG. The Mandell's It's Not Your Fault You're Fat Diet. New York, NY: Harper and Row, 1983.

91. Reiser S, Ferretti RJ, Fields M, Smith JC Jr. Role of dietary fructose in the enhancement of mortality and biochemical changes associated with copper deficiency in rats. Am J Clin Nutr 1983; 38:214-22.

92. Fields M, Michaels OE IV, et al. Effect of copper deficiency on intestinal hexose uptake and hepatic enzyme activity in the rat. Nutr Rep Int 1983; 28:123-31.

93. Rayssiguier Y, Gueux E, Weiser D. Effect of magnesium deficiency on lipid metabolism in rats fed a high carbohydrate diet. J Nutr 1981; 111:1876-83.

94. Fields M, Reiser S, Smith JC Jr. Effect of copper and zinc on insulin binding and glucose transport by isolated rat adipocytes. Nutr Rept Int 1983; 28:163-69.

95. Thornber JM, Eckert CD. Protection against sucrose-induced retinal capillary damage in the Wistar rat. J Nutr 1984; 114:1070-75.

96. Copper: What a difference sex makes. Science News 1987; 131:70.

97. Khaw K-T, Barrett-Conner E. Dietary potassium and blood pressure in a population. Am J Clin Nutr 1984; 39:963-68.

98. Kromhout D, Bosschieter EB, Coulander C. Potassium, calcium, alcohol intake and blood pressure: the Zutphen study. Am J Clin Nutr 1985; 41:1299-1304.

99. McCarron DA. Calcium and magnesium nutrition in human hypertension. Ann Intern Med 1983; 98(Part 2):800-805.

100. Langford HB. Dietary potassium and hypertension: epidemiologic data. Ann Intern Med 1983; 98(Part 2):770-72.

101. Tannen RL. Effects of potassium on blood pressure control. Ann Intern Med 1983; 98(Part 2)773-80.

102. Dyckner T, Wester PO. Effect of magnesium on blood pressure. Br Med J 1983; 286:1847-49.

103. MacGregor GA, Smith SJ, et al. Moderate potassium supplementation in essential hypertension. Lancet September 11 1982; pp. 567-70.

104. Khaw K-T, Thom S. Randomized double-blind cross-over trial of potassium on blood-pressure in normal subjects. Lancet November 20, 1982: pp. 1127-29.

105. Iimura O, Kijima T, et al. Studies on the hypotensive effect of high potassium intake in patients with essential hypertension. Clin Sci 1981; 61:77s-80s.

106. Khaw K-T, Barrett-Conner E. Dietary potassium and stroke-associated mortality. N Eng J Med 1987; 316:235-40.

107. Gonzalez-Calvin JL, et al. Efectos de la ingestion de sacarosa sobre la diuresis, calciuria y otros constituyentes urinarios en sujectos sanos. Rev Clin Esp 1981; 160:293-97.

108. Boldin BR, Swenson L, Dwyer J, Sexton M, Gorbach L. Effect of diet and Lactobacillus acidophilus supplements on human fecal bacterial enzymes. JNCI 1980; 64:255-61.

109. Bilsing L. Use nature's methods to control pests. Environ Fall 1985; 2:8-9.

110. Raghunandana RS, Srinivasa R, Venkataraman PR. Investigations on plant antibiotics. J Scien Indust Res August, 1946; 1B:31-35.

111. Bruggeman IM, Temmink FHM, Van Bladeren PJ. Glutathione and cysteine-mediated cytotoxicity of allyl and benzyl isothiocyanate. Toxicol Appl Pharmacol 1986; 83:349-59.

112. Appleton JA, Tansey MR. Inhibition of growth of zoopathogenic fungi by garlic extract. Mycologia 1975; 67:881-85.

113. Barone FE, Tansey MR. Isolation, purification, identification, synthesis, and kinetics of activity of the anticandidal component of Allium sativum, and a hypothesis for its mode of action. Mycologia 1977; 69:792-825.

114. Hasilik A. Perturbation of growth and metabolism in Candida albicans by 4-bromobenzylisothiocyanate and iodoacetate. Naturforsch Jan-Feb 1973; 28:21-31.

115. Truss CO. The Missing Diagnosis. Birmingham, AL: The Missing Diagnosis, Inc., 1983.

116. Crook WG. The Yeast Connection. Jackson, TN: Professional Books, 1983.

117. Loranzani SS. Candida: A Twentieth Century Disease. New Canaan, CT: Keats Publishing Inc., 1985.

118. Wunderlich RC Jr, Kalita DK. Candida Albicans: How to Fight an Exploding Epidemic of Yeast-Related Diseases. New Canaan, CT: Keats Publishing, 1986.

119. Bantle JP, Laine DC, et al. Postprandial glucose and insulin responses to meals containing different carbohydrates in normal and diabetic subjects. N Eng J Med 1983; 309:7-12.

120. Glinsmann WH, Irausquin H, Park YK. Evaluation of health aspects of sugars contained in carbohydrate sweeteners: report of sugars task force 1986. J Nutrition Supplement November, 1986. Vol. 116 Number 11S.

121. Glinsmann WH, Irausquin H, Park YK. Evaluation of health aspects of sugars contained in carbohydrate sweeteners: report of sugars task force, 1986. J Nutr 1986; 116(11S):S97,S115-17.

122. Srinivasan SR, et al. Synergistic effects of dietary carbohydrate and cholesterol on serum lipids and lipoprotiens in squirrel and spider monkeys. Am J Clin Nutr 1978; 31:603-13.

Chapter Seven

1. Gilbert RM. Caffeine: overview and anthology. In: Nutrition and Behavior. Ed: Miller SA. Philadelphia, PA: The Franklin Institute Press, 1981.

2. Gilbert RM, Marshman JA, Schwieder M, et al. Caffeine content of beverages as consumed. Can Med Assoc J 1976; 114:205-08.

3. Martinek RG, Wolman W. Xanthines, tannins and sodium in coffee, tea and cocoa. JAMA 1955; 158:1030-31.

4. Bunker ML, McWilliams M. Caffeine content of common beverages. J Am Diet Assoc 1979; 74:30.

5. Costill DL, Dalsky GP, Fink WJ. Effects of caffeine ingestion on metabolism and exercise performance. Med Sci Sports 1978; 10:155-58.

6. Ivy JL, Costill DL, Fink WJ. Influence of caffeine and carbohydrate feedings on endurance performance. Med Sci Sports 1979; 11:6-11.

7. Smith DL, Tong JE, Leigh G. Combined effects of tobacco and caffeine on the components of choice reaction-time, heart rate, and hand steadiness. Percept Mot Skills 1977; 45:635-39.

8. Stillner V, Popkin MK, Pierce CM. Caffeine-induced delirium during prolonged competitive stress. Am J Psychiatry 1978; 135:855-56.

9. Eysenck MW, Folkard S. Personality, time of day, and caffeine: some theoretical and conceptual problems in Revell et al. J Exp Psychol 1980; 109:41-43.

10. Revelle W, Humphreys MS, Simon L, et al. The interactive effect of personality, time of day, and caffeine: a test of the arousal model. J Exp Psychology 1980; 109:1-31.

11. Wright LF, Gibson RG, Hirshowitz RI. Lack of caffeine stimulation of gastrin release in man. Proc Soc Exp Biol Med 1977; 154:538-39.

12. Cohen S, Booth GH. Gastric-acid secretion and the lower-esophageal-sphincter pressure in response to coffee and caffeine. N Eng J Med 1975; 293:897-99.

13. Truitt EB Jr. The Xanthines. In: Drill's Pharmacology in Medicine 3rd Edition. Ed: DiPalma JR. New York, NY: McGraw-Hill Book Company, 1965; pp. 394-409.

14. Estler CJ. Failure of phenoxybenzamine and pimazide to diminish changes in oxygen consumption and body temperature produced by caffeine. Arch Int Pharm Ther 1979; 239:326-30.

15. Callahan MM, Rohovsky MW, Robertson RS, et al. The effect of coffee consumption on plasma lipids, lipoproteins, and the development of aortic atherosclerosis in Rhesus monkeys fed an atherogenic diet. Am J Clin Nutr 1979; 32:834-45.

16. Ax RL, Collier RJ, Lodge JR. Effects of dietary caffeine on the testis of the domestic fowl, Gallus domesticus. J Reprod Fertil 1976; 47:235-38.

17. Ax RL, Lodge JR. Effects of dietary caffeine on the reproductive performance of chickens. Ill Res 1976; 18:18-19.

18. Friedman L, Weinberger MA, Farber TM, et al. Testicular atrophy and impaired spermatogenesis in rats fed high levels of the methylxanthines caffeine, theobromine, or theophylline. J Environ Pathol Toxicol 1978; 2:687-706.

19. Terada M, Nishimura H. Mitigation of caffeine-induced teratogenicity in mice prior chronic caffeine injection. Teratology 1975; 12:79-82.

20. Weathersbee PS, Lodge JR. Caffeine: its direct and indirect influence on reproduction. J Reprod Med Obstet Gynecol 1977; 19:55-63.

21. Weathersbee PS, Olsen LK, Lodge JR. Caffeine and pregnancy. Postgrad Med 1977; 62:64-69.

22. Streissguth AP, Martin DC, BArr HM. Caffeine effects on pregnancy outcome. Unpublished manuscript, University of Washington, Seattle.

23. Jacobson MF. Caffeine poses risk to the fetus - FDA criticized for inaction. Brief submitted to the US Food and Drug Administration by the Center for Science in the Public Interest, August 1978.

24. Hennekens CH, Drolette ME, Jesse JM, et al. Coffee drinking and death due to coronary heart disease. N Eng J Med 1976; 294:633-36.

25. Heyden S, Heyden F, Heiss G, et al. Smoking and coffee consumption in 3 groups: cancer deaths, cardiovascular deaths and living controls. A prospective study in Evans County, Georgia. J Chronic Dis 1979; 32:673-77.

26. Rotenberg FA, DeFeo JJ, Swonger AK. Coffee, myocardial infarction, and sudden death. Lancet 1979; 2:140-41.

27. Coffee consumption tied to risk of heart disease. AMA News November 22/29, 1985.

28. Williams PT, Wood PD, et al. Coffee intake and elevated cholesterol and apolipoprotein in men. JAMA 1985; 253:1407-11.

29. Little JA, Shanoff HM, Csima A, et al. Coffee and serum-lipids in coronary heart-disease. Lancet 1966; 1:732-34.

30. Thelle DS, Arnesen E, Forde OH. The Tromso heart study: does coffee raise serum cholesterol? N Eng J Med 1983; 308:1454-57.

31. Young G. Chocolate: food of the gods. National Geographic November 1984; pp. 664-87.

32. Composition of Foods. United States Department of Agriculture. Agriculture Handbook Number 8.

33. Sclafani A. Dietary Obesity. In: Obesity. Ed: Stunkard AJ. Philadelphia, PA: W.B. Saunders Company, 1980; pp. 166-81.

Chapter Eight

1. Mandell M, Mandell FG. The Mandell's It's Not Your Fault You're Fat Diet. New York, NY: Harper and Row, 1983.

2. Berger SM. Dr. Berger's Immune Power Diet. New York, NY: NAL Books, 1985.

3. Levin AS, Zellerbach M. The Type 1/Type 2 Allergy Relief Program. Los Angeles, CA: Jeremy P Tarcher, Inc. Distributed by Houghton Mifflin Company, Boston 1983; pp. 119-33.

Chapter 10

1. Nakamura MM, Overall JE, Hollister LE, Radcliffe E. Factors affecting outome of depressive symptoms in alcoholics. Alcoholism (NY) Spring 1983; 7/2:188-89.

INDEX

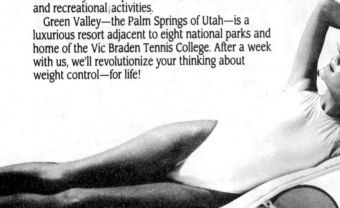

How To Lower Your Fat Thermostat

Diets don't work and you know it! The less you eat, the more your body clings to its fat stores. There is only one program that teaches you to eat to lose weight and it's detailed here in this nationwide best-selling book. All other weight-control programs are based on caloric deprivation. The *How To Lower Your Fat Thermostat* program is based on giving you enough total calories and nutrients to convince the control centers in your brain that regulate fat stores that you don't need to hold onto that fat any more. Then your weight will come down naturally and comfortably, and stay at that lower level permanently.

Recipes To Lower Your Fat Thermostat

Companion cookbook to *How To Lower Your Fat Thermostat*. Once you understand the principles of the fat thermostat program, you will want to put them to work in your daily diet. Now you can with this full-color, beautifully illustrated cookbook. New ways to prepare more than 400 of your favorite recipes. Breakfast ideas. Soups and salads. Meats and vegetables. Wok food, potatoes, beans, and breads. Desserts and treats. All designed to please and satisfy while lowering your fat thermostat.

Acrylic Cookbook Holder

This acrylic cookbook holder is the perfect companion to your new cookbook. Designed to hold any cookbook open without breaking the binding, it allows you to read recipes without distortion while protecting pages from splashes and spills.

Five Roadblocks to Weight Loss (Audiocassette)

If you have a serious weight problem that has failed to respond to the fat thermostat program, then you could be suffering from any of the five roadblocks to weight loss: food addictions, artificial sweeteners, food allergies, yeast overgrowth, and stress. Learn what these roadblocks are, what to do about them, and how the fat thermostat program relates to them . . . in an exclusive interview with Drs. Dennis Remington and Edward Parent.

Pocket Progress Guide

A pocket-sized summary of the fat thermostat program that includes food composition tables, daily records, and a progress summary for quick and easy reference and record-keeping anytime, anywhere.

The Neuropsychology of Weight Control
(8 Audiocassettes and Study Guide)

Based on the best-selling book, *How To Lower Your Fat Thermostat*, this audiocassette program explains the principles of the fat thermostat program, then teaches you how to reprogram your fat thermostat for leanness. You will learn how to take control of your body and mind, how to determine your ideal body image, how to develop a fat-burning mechanism in the brain, and—best of all—how to develop a lifelong blueprint for leanness and health.

The Neuropsychology of Weight Control
(Videocassette)

For some people, seeing is believing. While reviewing the key points of the program and the benefits of reaching your goal weight, this motivational video also features testimonials by people who have had dramatic success. In moments of doubt or discouragement, this video provides the needed support and encouragement.

Computerized Diet Analysis

Even when you try to eat right, your diet could be missing calories and nutrients vital to good health and weight control. One of the best ways to determine whether you are getting all the nutrients you need is to analyze your actual food intake for several days. Our *Computerized Diet Analysis* provides you with a complete computer print-out detailing the calories, fat, protein, carbohydrates, and 29 different vitamins, minerals, and other essential nutrients present in your current diet. Along with the print-out, you will receive a computer graph that indicates the percentages of the Recommended Daily Allowances (RDA) that your current diet provides. In addition to that, you will receive a brief written summary of your diet listing the areas that need attention along with suggestions for improvement—all prepared by a registered dietitian.

Instructions: Eat as you normally would (without trying to make any improvements), recording by date your food intake for two weekdays and one weekend (four days total). In recording, note the amounts of each food item ingested in teaspoons, tablespoons, ounces, cups, number of pieces, etc. (You may have to estimate.) Include in your record toppings, cooking method, and all other information vital to evaluating the food (i.e., chicken—white or dark meat, with or without skin, deep-fat fried or broiled; bread—white or wheat; milk—whole, 2%, 1% or skim). Using the order form provided, send your food record along with your name, address, sex, age, height and weight to Vitality House for your own personal computerized diet analysis.

The Bitter Truth About Artificial Sweeteners

Research proves that those people using artificial sweeteners tend to gain more weight. Not only do artificial sweeteners enhance the desire for sweets, they also cause many unpleasant side effects in addition to raising the fat thermostat. Learn the real truth about artificial sweeteners and sugars. Learn how they affect your health and weight and what you can do about them.

Back To Health: A Comprehensive Medical and Nutritional Yeast-Control Program

If you suffer from anxiety, depression, memory loss, heartburn or gas . . . if you crave sugar, chocolate or alcohol . . . if weight control is a constant battle . . . if you are tired, weak and sore all over . . . this book was written for you. While yeast occurs naturally in the body, when out of control it becomes the body's enemy, manifesting itself in dozens of symptoms. Getting yeast back under control can correct many conditions once considered chronic. More than 100 yeast-free recipes, plus special sections on weight control, hypoglycemia and PMS.

Order now and save $1.50 per item!

Qty.	Code	Description	Retail	Savings	Cost	Subtotal
	A	How To Lower Your Fat Thermostat	$ 9.95	$1.50	$ 8.45	
	B	Recipes To Lower Your Fat Thermostat	$14.95	$1.50	$13.45	
	C	Acrylic Cookbook Holder	$ 9.95	$1.50	$ 8.45	
	D	Neuropsychology of Weight Control (8 Audiocassettes and Study Guide)	$69.95	*	$69.95	
	E	Back To Health	$ 9.95	$1.50	$ 8.45	
	G	Bitter Truth About Artificial Sweeteners	$ 9.95	$1.50	$ 8.45	
	H	Five Roadblocks to Weight Loss (Audiocassette)	$ 7.95	$1.00	$ 6.95	
	I	Pocket Progress Guide	$ 3.00	$ .50	$ 2.50	
	J	Neuropsychology of Weight Control (Videocassette)	$29.95	*	$29.95	
	K	Computerized Diet Analysis (Include Food Intake Forms with order)	$25.00	$1.50	$23.50	
	Z	Green Valley Health Resort Information Packet	FREE	FREE	FREE	
Shipping and handling $1.50 per item						
2 or more items, we pay shipping and handling						N/C
Utah residents add 6.25% sales tax						
TOTAL						

* Order includes FREE any book up to $14.95 in value.

☐ Check ☐ Money Order ☐ MasterCard ☐ VISA ☐ American Express

Card No. _____ Expires _____

Signature _____ Phone _____

Ship to: Name _____

Address _____

City/State/Zip _____

Mail to: Vitality House International
3707 N. Canyon Rd./8-C, Provo, UT 84604 (801) 224-9214